ONTARIO WEATHER

Catastrophes, Ice Storms, Floods, Tornadoes and Hurricanes

A.H. Jackson

BLUE BIKE BOOKS

© 2009 by Blue Bike Books
First printed in 2009 10 9 8 7 6 5 4 3 2 1
Printed in Canada

All rights reserved. No part of this work covered by the copyrights hereon may be reproduced or used in any form or by any means—graphic, electronic or mechanical—without the prior written permission of the publisher, except for reviewers, who may quote brief passages. Any request for photocopying, recording, taping or storage on information retrieval systems of any part of this work shall be directed in writing to the publisher.

The Publisher: Blue Bike Books
Website: www.bluebikebooks.com

Library and Archives Canada Cataloguing in Publication

Jackson, A. H., 1944–
  Weird Ontario weather: Catastrophes, Ice Storms, Floods, Tornadoes and Hurricanes / Alan H. Jackson.

Includes bibliographical references.
ISBN 978-1-897278-48-2

  1. Ontario—Climate—Miscellanea. 2. Weather—Miscellanea. I. Title.
QC985.5.O6J32 2009   551.65713   C2009-902759-3

Project Director: Nicholle Carrière
Project Editors: Paul Deans, Pat Price
Cover Image: ©Jacysworld | Dreamstime.com; ©Geckophoto | Dreamstime.com
Back Cover: Photos.com
Illustrations: Peter Tyler, Roger Garcia, Roly Wood
Photo credits: Photographs courtesy of © Ba-Mi | Dreamstime.com (p. 122); © Cringuette | Dreamstime.com (p. 25); © Gareth12468 | Dreamstime.com (p. 74); © Geckophoto |Dreamstime.com (p. 99); © Jakeblaster | Dreamstime.com (p. 47); © Jschmidty | Dreamstime.com (p. 65); © Mollypix | Dreamstime.com (p. 174); © Wowbagger | Dreamstime.com (p. 19).

We acknowledge the support of the Alberta Foundation for the Arts for our publishing program.

We acknowledge the financial support of the Government of Canada through the

Canadian Heritage   Patrimoine canadien

Book Publishing Industry Development Program (BPIDP) for our publishing activities.

DEDICATION

For Maria—who makes the sun shine.

–A.H. Jackson

ACKNOWLEDGEMENTS

Freeman Dyson is a man who keeps his wits while his peers succumb to confused prophesies. Scientist, philosopher and intellectual, Dyson is welcome ice in a scientific hot pot, and for that I am grateful. Writing books on weather, especially weird weather, requires research on subjects not usually entertained and as diverse as tornadic velocity and potatoes.

The potatoes came from Dyson—a hammer stroke, as I had never considered the lowly spud as being anything more than a side for a good steak. Who would have thought the potato might be the saviour of civilization? Deep stuff, which is what spuds have—roots deep enough to suck frozen water from Martian soil. Imagine: instead of Yukon Golds, future space travellers will have Martian Reds with their soy burgers. Local produce will be an important consideration should the big blue become a big ball of brown sludge and force mankind to emigrate.

A mass exodus from terra firma is something I had considered unfeasible until Dyson pointed out its inevitability. Cheap is the key, and when that day arrives (and it will), millions of folks will bolt into space accompanied by their designer dogs and cats. Makes you think of Ireland—all those Martian settlers depending on the spud and suddenly being let down by potato blight and having to pack up and find other shores. A grand design perhaps—interstellar settlement by the human race. Who knows?

Nobody does, for sure, which is how Dyson views present-day climate change. Nothing is new. Climate now has been before and will be again; climatologists and green gangers should wait and see what happens before pushing panic buttons. They have been pushing a lot of buttons, and kudos to Freeman Dyson for ridiculing their dependence on computer models that ignore biology and the general spunk of the big blue.

Also, kudos to Environment Canada for keeping track of climate happenings and not pontificating results like hippy dippy weathermen. Those guys are pros, and I am grateful to them for keeping reminders of all I have forgotten about climate and weather.

Still more kudos to the folks at Blue Bike Books for allowing a gentler voice to be heard through the racket created by legions of doomsayers and bowling-for-dollars ecology gangs. Those that can, do; and those that can't will teach you until you're out of money faster than you can say climate change.

CONTENTS

INTRODUCTION 8
Climate vs. Weather 9
Climate Change 15

CHAPTER 1: WEATHER: A FEW BASICS 17
We've Got Atmosphere 18
Atmospheric Highways 27

CHAPTER 2: A WEATHERY HISTORY OF ONTARIO 30
Weather Shapes the Province 31
A Climatic Explanation 37

CHAPTER 3: ONTARIO WEATHER TODAY 42
Where Does it Come From? 43

CHAPTER 4: ONTARIO'S WEIRD WEATHER PLACES 48
The North 49
The Great Lakes 53
Odd Landforms 74
Eastern Ontario 78

CHAPTER 5: SIGNIFICANT ONTARIO WEATHER EVENTS . 84
Ontario Weather Disasters 85
Surviving Natural Disasters 96

CHAPTER 6: TORNADOES 98
Doing the Twist(er) 99
Safety First 110

CHAPTER 7: LIGHTNING 113
A Charged Affair 114
Be Lightning Wise 121

CHAPTER 8: FLOODS 125
Soggy Situations 126
Assume the Worst 130

CHAPTER 9: HAIL 135
Cannonballs from the Sky 136

CHAPTER 10: LANDSLIDES 139
Slip Sliding Away. 140

CHAPTER 11: SMOG 143
Out with the Bad Air 144

CHAPTER 12: FOREST FIRES AND INSECTS 147
Burns and Bugs. 148

CHAPTER 13: WINTER'S NASTY BITE 152
Blizzards and Cold 153
Frostbite and Hypothermia. 156
Giant Snowflakes. 159

CHAPTER 14: WEATHER MODIFICATION 161
Weather Wizards 162
Hail Suppression 167

CHAPTER 15: ATMOSPHERIC OPTICS 169
Look Up, or Miss the Show. 170

CHAPTER 16: ONTARIO WEATHER STORIES 177
Hot Times in Timmins. 178
The Angel of Long Point. 184
Hat Tricks 190
The Rescue. 193
A Bible Story Redux 201

CHAPTER 17: CONCLUSIONS 203
Weather Thoughts. 204
Climate-change Thoughts. 208

NOTES ON SOURCES 212

Introduction

Climate is what we expect, weather is what we get.

–Mark Twain

INTRODUCTION

CLIMATE VS. WEATHER

You Say Climatology, I Say Meteorology

What is the difference between weather and climate? Weather is simply a disruption of the long-term expectation of what we call the climate. Residents of southern Ontario expect warm summers interspersed with periods of cooler temperatures and storms—that's climate. In July 1936, they got 10 days of unrelenting heat that topped 42°C—that's weather, sizzling hot weather.

Weather and climate are two distinct fields of study. Meteorologists concern themselves with weather occurring now, while climatologists work with relics of past events: tree rings, sediment cores from ocean bottoms, glaciers, pollen trapped in ice, amber and sedimentary strata in rocks. While divergent in studies, both these schools work toward the same goal: a better understanding of weather events.

> *According to a news story, if global warming continues, in 20 years the only chance we will have to see a polar bear is in a zoo. In other words, nothing is going to change.*
>
> —Arthur Carlson, fictional character and station manager of *WKRP in Cincinnati*

Is Ontario's weather and climate changing? Sure it is. Ever since the beginning of time, no two days, weeks, months or years have seen exactly the same weather in exactly the same place. Our world is constantly in flux—volcanoes erupt, forests burn, ocean currents reposition and climates change.

Not to worry; what is happening now has happened before and might be a good thing for Ontario, because some scientists believe the present warming trend is forestalling another mini ice-age episode. Ontario may stand to benefit from a warming climate. An increase in the amount of carbon dioxide in our atmosphere could allow for faster growth of forests and grain crops, while open seas in the north could facilitate shipping out the bounty. Nobody knows what global warming means for Ontario, but it certainly will not be as disastrous as the media doomsayers claim. Be leery of doomsayers, green gangs and seven-day weather forecasters.

Predicting the Weather

There once was a popular weather forecaster who locked himself in his office before every show to consult a mysterious black box. One day, the weather forecaster died, and his staff rushed in to have a look inside the mysterious black box. Inside they found a paper with the words: "high-pressure = good weather, low-pressure = bad weather."

A joke, but not far off the mark, because weather forecasting for small radio and TV stations used to be the starting job for those wishing a career in broadcasting. I can remember hearing a local radio weather forecaster predict intermittent rain just before Hurricane Hazel struck in 1954. A bad call, but it probably never bothered that forecaster, because the job has never depended on being right—it depends on presence.

A weather prognosticator must look great and have a good act. Being a weatherperson is not a hard job, but becoming a professional meteorologist is—it requires years of university study and training. Unfortunately, few small stations can afford the services of a professional, and local residents must rely on predictions from a hippy dippy weatherman familiar with local weather probabilities. In most areas of Ontario,

INTRODUCTION

approximately 16 percent of all days see precipitation, so if a weathercaster calls for no precipitation every day, he or she will be right approximately 84 percent of the time. Can you see the problem?

Most weathercasters do "now" weather—six-hour predictions—and because of satellite feeds and connections with Environment Canada, they get it right most of the time. But problems arise with their down-the-road predictions, the seven-day forecast you need to plan your week. They are astonishingly bad at forecasting for long periods, and this is where a good act comes in handy—along with a scapegoat. You played golf in pouring rain instead of sunshine, and the weathercaster you want to strangle is on the tube blaming climate change.

INTRODUCTION

Not just weathercasters but hoards of media people have leapt onto that bandwagon until is seems that every third program on television or radio is about climate change, a subject with many experts and pontificators. Mostly they spout nonsense, because there is nothing new under the sun; weird weather happening now has happened before and will happen again. But even weird weather needs perspective. How does our sense of weird stack up to, let's say…deep space, where radio astronomers recently found acetic acid (that's right, vinegar) in a huge cloud of gas called Sagittarius B2 North. And not to be outdone on the weirdness scale, British astronomers recently identified a cloud composed mostly of alcohol, enough to fill a half billion swimming pools. Now that…is weird.

Can't a Planet Get a Little Respect?

Almost as strange is the total lack of respect humanity has shown our planet; we take too much, give back little, think climate is subservient and act with surprised indignation when it undergoes a change. Are we to blame? Are we polluting the world and causing climate change? Sure. Our world is over-populated, over-urbanized, insanely industrialized and careless with toxic wastes.

How we treat this planet affects Ontario's weather. That old adage, "you reap what you sow," is on the button but old-fashioned. A more modern version should read, "what you reap may come back to haunt you." Civilization has reaped a lot: 15 million square kilometres plowed up for farmland, 10 million square kilometres of forest cut down, and 32 million square kilometres set aside for livestock grazing. That leaves only 90 million square kilometres, and loggers are attacking that vigorously.

Deforestation is progressing so rapidly that, just to keep pace with loss and allow for slow regeneration, 250 million trees

INTRODUCTION

would have to be planted every year—and we should be doing that because trees take in carbon dioxide and modify temperature extremes. Trees also keep soil from eroding, and we need that because, every year, around 75 million tons of precious topsoil is blown away or eroded by rainwater. In 1776, the average depth of topsoil on U.S. and Canadian prairie lands was 22 centimetres, a mean depth that over the years has diminished to less than 15 centimetres. Not good, and it's worse in some undeveloped countries with unfettered population growth, where soil depth is only half what it was a century ago. A grim future may await those nations, as sometime down the road, their growing populations could experience starvation on a massive scale.

During the 1800s, lumber barons and immigrant farmers laid waste to the forests of southern Ontario, resulting in the loss of topsoil by wind and rain erosion over vast areas. Some regions have seen reforestation, but many depend on sandy soil crops such as tobacco, corn and soybeans to keep land from becoming bedrock. Counties like Essex and Kent, with their vast fields of corn and soybeans are prone to weird weather, as flat land is a magnet for supercell storms. Those and adjacent counties see extraordinary thunderstorm and tornado activity in summer and blizzard conditions in winter.

Residents of northern Ontario are accustomed to flatland weather but can experience drastic changes because of forest fires; no trees to absorb sunlight means reflected radiation and disrupted rainfall. Local residents blame it on climate change, when the actual cause is the fire-ravaged slopes of nearby hills. Before the fire, air cooled and lost water vapour as it flowed up the slopes. After a fire, the rising air warms, and clouds fail to produce precipitation, or it rains farther down the line. A change in normal rainfall patterns means drier forests and more fires. Northern Ontario is a virtual snakes-and-ladders

INTRODUCTION

weather game: the government concentrates on reforesting one area, while elsewhere the stage is set for another blaze. The Ontario government does a good job controlling fires, but reforestation depends on budgets, and, though officials can anticipate hot spots, they're powerless to do much more than warn off campers.

A good number of those conditions the media Moses people call "contributors to global climate change" are actually symptoms of a planet made sick by overpopulation. Stripping the planet of protective forests and groundcover causes greenhouse gas emissions to rise, even without burning fossil fuels. Magazines used to have great full-page cartoons, and the one I remember best featured two Martians approaching Earth in a flying saucer: one, with a horrified look on its face, said, "Don't land! That planet is infected with people."

INTRODUCTION

CLIMATE CHANGE

We Are Obsessed

The fact that climate is warming doesn't scare me a bit. All the fuss about global warming is grossly exaggerated. The polar bears will be fine.

–Freeman Dyson, physicist

Not your run-of-the-mill physicist, Freeman Dyson is up there on the brain scale with Albert Einstein and thinks climate doomsayers are guilty of spouting "lousy science" that distracts from more immediate problems. Problems such as nuclear weapons, toxic pollution and mass starvation through overpopulation. He thinks the Green Movement has become a bandwagon for political, scientific and financial opportunists who employ "if you're not with us, you're a bad person" tactics to bully the public into accepting their own personal views of environmental situations. He contends that there are problems, and agrees with the prevailing view that rapidly rising carbon-dioxide levels in the atmosphere caused by human activity exist, but things are not nearly as bad as the green gangs would like us to believe.

Dyson believes if global warming melts polar ice and causes sea levels to rise, we will simply have to adapt. Dyson likes coal, an energy commodity the Greens bedevil constantly but is so cheap most of the world can afford it. He thinks such a plentiful supply of cheap energy is paramount for the economic advancement of developing countries; especially when the technology exists to scrub it clean of carbon. China and India will never stop burning coal, but they may be agreeable to installing advanced clean-coal technology in their generating plants. Grind coal into dust, mix with water and you have a fuel that burns with the efficiency of natural gas, and that's probably the future of coal.

Dyson thinks that the initial stages of evolution occurred on our planet during a time when our atmosphere contained high levels of carbon dioxide—much higher than today's concentrations. He believes climate change has become an obsession—the primary article of faith for a worldwide secular religion called sentimentalism. Greens make assumptions based on their climate models that take into account atmospheric motion and water levels but have no feeling for the chemistry and biology of sky, soil and trees. "The biologists have essentially been pushed aside," says Dyson, who feels that if the going gets tough, the tough will get going—and do things like plant a billion trees, stop using the oceans as garbage dumps and curtail our raging population growth.

Weather: A Few Basics

*It was a wide space; I could tell you how wide,
in chains, perches, furlongs, and things,
but that would not help you any.*

–Mark Twain

WE'VE GOT ATMOSPHERE

Weather Wisdom

All books meant to inform usually contain chapters wherein the author attempts to impart necessary subject information by the "tap, tap, listen up, class," method. A lecture that invariably leads to mind-numbing boredom and sleepy time. This is no sleepy-time book; somewhere between its covers you might find and remember a few life-preserver sentences that could save your hide from a weird weather event.

So—tap, tap, listen up, class! Weird weather is dangerous and to understand how and why needs some basic grounding. It's only the one chapter, so take a deep breath and begin. It'll be over soon and will leave you transformed.

Cloud Primer

Clouds are either cumulus or stratus; the former is puffed up by rising warm air, the latter is a high-altitude fog created by air that doesn't rise. Clouds are further divided into four altitudes: towering, high, middle and low.

Towering clouds can rise to more than 25,000 metres and behave badly; these black hats often have the Latin word *nimbus* (rain cloud) tacked on so you know not to walk the dog. If you see one of these in the neighbourhood, stay inside.

High clouds, or alto, are composed of tiny ice crystals and found at altitudes of 6000 to 9000 metres. These are divided into three types: cirrus, sometimes called "mare's tails," that look like wispy streaks of white in a blue sky; cirrocumulus, sometimes called a "mackerel sky," which look rippled and wavy; and cirrostratus—thin, high-altitude gossamer sheets composed of ice crystals that are responsible for halos around the sun or moon.

Middle clouds are divided into alto or cumulus and found at altitudes between 2000 and 6000 metres. Altostratus are dense sheets of grey but can appear as stripes. The sun or moon is visible but looks hazy. Altocumulus are composed of water droplets and resemble mare's tails but are large and puffy. The sun shines through, but often has a corona that can be in various colours.

Low clouds begin at around 1800 metres and have three types: stratus, nimbo and strato.

Stratus is a thick grey fog covering the entire sky and is responsible for drizzling rain. These clouds form when there is little or no vertical movement of air.

Nimbostratus are rain clouds. These are dark, ominous and sometimes touch the ground. Nimbostrati are actually middle clouds with low bases and occurring precipitation but can belong to any grouping because their vertical extension can be massive.

Stratocumulus are irregular masses of various shades of grey spread out in layers. No rain from these clouds, but they bear watching because they sometimes fuse into nimbostratus and cause sudden downpours.

Cumulonimbus are the ones to watch. Bases may touch the ground and updrafts rise to more than 25,000 metres. High winds aloft can sometimes cleave their tops so they resemble a blacksmith's anvil. Thunderstorms and tornadoes spawn from these clouds as do positive lightning strikes, so don't walk the dog when clouds look threatening.

Cumulus are fine-weather clouds, unless they come together and form a cumulonimbus. Luckily, that does not happen often, since the average lifespan of a cumulus cloud is only 15 minutes.

Here is a list of cloud types preceded by their metrological designation letters—handy for reading online weather maps.

Cloud Types

AC—altocumulus
ACC—altocumulus castellanus
AS—altostratus
CC—cirrocumulus
CS—cirrostratus
CI—cirrus
CB—cumulonimbus
CU—cumulus
CF—cumulus fractus
SF—stratus fractus
TCU—towering cumulus
NS—nimbostratus
SC—stratocumulus
ST—stratus
F—fog
R—rain
A—hail
IP—ice pellets (including ice pellet showers)
L—drizzle (including freezing drizzle)
IC—ice crystals
S—snow (snow showers, snow pellets and snow grains)
BS—blowing snow
D—dust, blowing dust or dust storm
H—haze
N—sand, blowing sand or sand storm
K—smoke
VA—volcanic ash

Atmosphere Primer

Our atmosphere consists of five layers, three of which have narrow boundaries: the troposphere and tropopause boundary; the stratosphere and stratopause boundary; the mesosphere and mesopause boundary; the thermosphere; and the exosphere.

The troposphere contains the stuff we breathe and is where all the weather comes from. It rises to about eight kilometres at the poles and 18 kilometres at the equator. This zone contains most of the atmospheric mass and almost all the water vapour. In the upper reaches of this zone are found Rossby waves, strong undulating winds moving east to west on a global scale—better known as the westerlies. The tropopause boundary is a temperature inversion lying atop the troposphere in a narrow broken band that loosely extends from pole to pole. This is the point where air ceases to cool with altitude and begins to warm. In this boundary between troposphere and stratosphere are born the jet streams.

The stratosphere extends upward from the troposphere to a height of roughly 50 kilometres above Earth's surface and is almost entirely devoid of weather. Jet pilots like it for that reason and because the air is thin and offers little resistance to the aircraft. When you see a jet contrail, know it's there because the plane's engines are adding weather to the void—moisture from the exhaust freezes and forms ice crystals that eventually float down into the troposphere. The stratopause, at a height of 50 to 55 kilometres, is the boundary between the stratosphere and mesosphere and is the point where temperatures rising with altitude reach maximums.

The mesosphere extends upward from the stratosphere to a height of roughly 85 kilometres above Earth's surface. It contains minute amounts of water vapour and gasses, temperatures can be –100°C at its top, and it's here that meteors burn up. At night when you observe a "shooting star," you're looking at the mesosphere. This is also where strange noctilucent clouds form; more on them later. Satellite pictures of our planet all portray a narrow, dark-blue band around its outer edge; that is the mesosphere.

The thermosphere, sometimes called the ionosphere, is nearly empty space. Gas particles found here are ionized from constant exposure to cosmic rays, meaning their atoms are stripped of protons and electrons. They have no pizzazz and drift in lifeless layers that extend upward for almost 1000 kilometres. It's because of these layers that we can send radio messages from beyond the horizon—they bounce off and reflect down. Higher-frequency waves, those used in FM radio and television transmissions, are not affected and will travel through the layers into space. Temperature extremes abound in this zone; daytime temperatures can range up to 2000°C. The thermosphere is where the International Space Station orbits and the northern lights dance.

The exosphere is up, up and away empty. Lonesome gas particles drifting here are as hot as the sun during day and almost absolute zero (–273°C) at night. Any gas molecules found here are so far apart they rarely collide and can attain tremendous speeds. Oxygen and helium molecules rising into the exosphere pick up so much speed they escape Earth's gravity and head into space.

Wind Primer

*Wind, the season-climate mixer,
in my Witches' Weather Primer.*

*Says, to make this Fall Elixir,
first you let the summer simmer.*

—Robert Frost, poet

Build it, and they will come—a motion picture cliché that pretty much describes how most local winds are caused. Altocumulus clouds build inside a low-pressure cell, and surrounding air rushes in as if being sucked into a vacuum cleaner. It builds and the air comes, and if the clouds are big, high and fast moving, it will come at gale-force speed.

A gale is a strong wind with speeds of 62 to 74 kilometres per hour. Gale is number 8 on the Beaufort Wind Scale (a 0-to-12 scale devised in 1805 by Rear-Admiral Sir Francis Beaufort to enable sea captains to describe wind effect). A modernized version is still in use today for describing what winds do at different speeds; our number 8, now called a fresh gale, causes moderate waves and spindrift (foam) to fly from the tops—on land it causes twigs to snap and cars to veer. Zero on the Beaufort scale is a flat sea with no wind, and 12 is a hurricane: huge waves, air filled with spindrift, sea completely white with driving spray and visibility greatly reduced. Old Admiral Beaufort must have scratched that number 12 into the logbook next to his last will and testament.

Local winds flow from areas of high pressure to low, and global winds are caused by solar heat-absorption differentials between climate zones and the rotation of the Earth. This rotation causes centrifugal force, which bends winds toward the west in the Northern Hemisphere and to the east in the Southern Hemisphere. Called the Coriolis effect, this centrifugal bending is responsible for the rotation of cyclones and tornadoes.

Differential heating also causes local winds, such as sea or lake breezes, where water continues to give off heat as the land cools, causing a rapid circulation of air. Mountain and valley breezes are caused by a similar heating differential: mountain

slopes get first light and more heat than valleys and pull up cooler air; in the afternoon, the roles are reversed, and air is drawn down by the sun-warmed valleys.

There are three primary atmospheric circulation cells: the Hadley, Ferrel and Polar cells. Hadley circulation is confined to the equatorial regions and produces tropical easterly winds above and below the equator that converge to form trade winds. Ferrel circulation is confined to the upper latitudes and produces prevailing westerly winds, while Polar circulation provides polar easterlies. Narrow global wind belts called jet streams meander between Big Kahuna circulation cells, and at very high altitudes.

ATMOSPHERIC HIGHWAYS

Jet Streams

Jet streams are east-to-west, high-speed ribbons of air originating in the tropopause. They are formed by the deflection of high-speed, upper-atmosphere winds caused by the rotation of the earth, or Coriolis acceleration—the effect that causes draining water to swirl. Jet streams travel at high altitudes, weaving and dipping like snakes. Ontario's main concern is with the most northerly of these winds—the polar jet stream, a fast-moving whip of frigid air that can create troughs of low pressure and ridges of high pressure even at high altitudes. Bad enough, but dipping closer to the ground it becomes a low-level jet stream and may cause severe weather mischief when encountering warm air masses coming from the southwest. Thunderstorms, tornadoes and ice storms can result from a low flyer—even worse is a split, with one side heading south and pressing up wet air from the Caribbean. That has happened a number of times, and on each there has been hell to pay—the last and biggest payment being the Great Ice Storm of 1998.

During summer, the average location of the polar jet stream over central Canada is about 50° north latitude. In winter, it arcs northward over BC, forms a ridge over the province and then turns sharply southward over the Great Plains of the U.S. It then plunges as far south as northern Texas before curving northeastward over the Mississippi River valley. Finally, it wends its way eastward, passes over New England or Atlantic Canada and leaves North America.

Below the polar stream, and approximately 13 kilometres above the subtropical high-pressure zone, another easterly blowing wind snake can be found—the subtropical jet stream. The reason for its formation is similar to the polar jet stream, but the subtropical stream is weaker because of the lower latitudinal temperature and pressure gradient.

Jet streams can also be isolated phenomena called jet streaks. Rising and sinking air found upstream and downstream from jet streaks can cause dangerous thunderstorm activity, and these are monitored closely by weather services.

Note: Daily jet stream monitoring maps can be found online and are a necessary consult for the amateur meteorologist.

Highs and Lows

Warm air holds water vapour and rises. Because it rises, its weight is reduced, and therefore some of the column of mercury in a barometer falls back into the reservoir bulb indicating the arrival of a low-pressure area.

Cold air is dry and dense. Its weight pushes mercury up the barometer column to indicate high pressure. Low pressure usually indicates bad weather; highs are associated with sunshine and fair weather. Air tends to flow from a high to a low. Low-pressure cells rise when encountering areas of high pressure, and at cooler altitudes, the low-pressure cells lose water vapour—it rains, it pours, and it can sometimes get downright nasty.

Big lows crossing the breadth of North America are called cyclonic storm systems because they pinwheel counterclockwise due to the Earth's rotation (known as the Coriolis effect). Cyclonic storms can bring along all manner of nasty weather, including tornadoes. That's bad, but without rain our crops would dry up, so we endure the odd cranky storm cell and hope it will go away quickly.

WEATHER: A FEW BASICS

Ontario gets many highs as well, and most originate from 30° to 35° north and south, the so-called "horse latitudes." The horse latitudes are the birthing ground of Canada's high-pressure maritime weather cells. This area lies under the subtropical ridge; a global belt of high pressure caused by falling, vapour-exhausted equatorial air—hot air rises, cools at altitude and releases all its water in the form of rain. Bone dry, this cool air falls onto the ocean's surface and gently moves off in a north-easterly direction to form trade winds, the prevailing pattern of easterly flowing tropical air and the perfect situation for old-time sailing ships.

But pity those ships that blundered into the downfall, because constant high pressure causes flat seas with little or no wind, an atmospheric trap for sailing ships that required wind to move. At times, those ships would stay becalmed for weeks; a situation some historians believe earned this calm weather belt its name. Rumour had it that ships carrying horses and running short of animal fodder tossed half the horses overboard to save the remaining animals. The story is probably not true—the name most certainly came from the slang words "dead horse money," meaning the monies advanced to sailors before a ship sailed—an advance worked off over the length of the voyage, which in the horse latitudes could be a long time.

Aside from affording calm winds and seas, horse latitudes are dry as toast and mostly responsible for the world's temperate zones and the great deserts such as the Sahara, Kalahari and the U.S. southwest. Fortunately, the horse latitudes send mainly pleasant weather to Ontario.

Congratulations: you have now completed the course and been transformed into a weatherperson. Remembering only half of what you've just read could land you a forecasting job at any local radio or TV station.

A Weathery History of Ontario

*Sunshine is delicious, rain is refreshing,
wind braces us up, snow is exhilarating;
there is really no such thing as bad weather,
only different kinds of good weather.*

–John Ruskin

WEATHER SHAPES THE PROVINCE

Welcome to Bug Land

Ontario is Canada's second-largest province, encompassing more than one million square kilometres—a hard-to-imagine figure that's better visualized through comparison: bigger than France, bigger than Spain and bigger than those two countries put together. An enormous piece of real estate, it's populated by a little more than 12 million people, a lesser population than a dozen of the world's major cities. So big, so few—why is that? Weird weather and bugs are the main reasons, and north of the transcontinental rail system, they're what people talk about—constantly. Too cold, too hot, too wet and pray for a snowfall in May to retard the black fly and mosquito hatch.

A WEATHERY HISTORY OF ONTARIO

Ontario stretches from the province of Manitoba in the west to Québec in the east, a distance of 2100 kilometres, and it encompasses all the area north from the American border to the shores of Hudson Bay, a distance of more than 1600 kilometres. Southern Ontario is a several-hundred-kilometre-wide strip of over-urbanized pasture and farmland bordering the Great Lakes and the U.S., while the north is a band of rocky, boreal forest edging a vast, almost deserted bog. The north runs from Lake Huron to Hudson Bay, an area comprising 87 percent of Ontario's landmass but only six percent of its population.

Ontario's near north is a vast boreal forest of spruce, fir and pine intertwined by a quarter-million lakes and rivers, while the far north up to the subarctic is an immense bog larger than most countries. Residents and tourists inhabit the near north and carnivorous bugs the boggy parts, a fortuitous division, unless warm weather explodes the insect populations and blows them south. Half-eaten tourists are counterproductive for that industry and best kept a secret. Only locals know the best repellent for black fly is a shotgun.

A WEATHERY HISTORY OF ONTARIO

A secret out there for all to see is in Upsala, a small Ontario town 140 kilometres west of Thunder Bay, where residents have erected a giant, roadside statue of a knife-and-fork-wielding mosquito poised to devour a hapless tourist. Tourists laugh at the giant bug, but you'd better not, because confronted by a wind-driven cloud of black flies, you'll pray for a shotgun with double barrels.

Years ago, while picnicking with friends on a smidgen of sand in Lake Erie called Ryerson's Island, I looked into a cloudless sky and noticed a singular black cloud. I thought it strange, and a few seconds later when I glanced up, the cloud was appreciably closer and making a noise. Black flies! That swarm attacked like piranhas and turned our 400-metre dash (in swimsuits) to our boat into an absolute horror as they bit us bloody stem to stern. Black flies can drive wild animals and people insane and kill them, but luckily their season is short—usually mid-June to mid-July. Unlucky is a weird weather heat wave in August and another hatching. Black flies are horrid little beasts, and it's difficult to tell when their season is over because of the hungry mosquitoes and deer flies. Depending on the weather, a summer in Ontario's far north can see you swaddled in mosquito netting and smelling like a chemical factory for weeks at a time.

Weather dictates all activities in Ontario's north. A heat wave will hatch billions of bugs and keep you trapped inside, while inversely, a cold snap can mean relief from bugs but still have you trapped inside for endless rounds of monopoly and jigsaw puzzles…all of which will leave you gaga and mindlessly watching beavers out the window.

The Beavers and the Bay

Beavers know nothing about games, spend most of their lives in water away from insects and live within a few metres of their bud 'n' bark food supply. The furry rodent has a good life in the north and cannot help but prosper. During the first half of the 17th century, when the Europeans began to sally into Ontario's far north by way of Hudson Bay, they stumbled onto a beaver bonanza that quickly fuelled a millinery revolution—the very English top hat. The race for beaver pelts was on, and thousands of Europeans arrived to trap the furry rodents and satisfy the London hat makers' insatiable demand. Most made money and left when area beaver populations waned.

Some trappers, probably driven insane by bugs, stayed on to trap other species. They soon found that wolf pelts made warm coats and bear fat could be rendered for grease to provide protection from insects. Native peoples around the Hudson Bay area thought European traders touched by spirits and treated them with reverence, an attitude that caught the attention of profit-minded London fur buyers who formed "The Governor and Company of Adventurers of England trading into Hudson's Bay."

On May 2, 1670, England's King Charles II gave the big swamp and everything west to the Rocky Mountains to the newly renamed Hudson's Bay Company and turned those touched fur traders into near-royalty. Armed with a royal charter, the company sent out representatives with orders to solicit more reverence (and pelts) from Native peoples in return for cheap trade goods. Company managers, or factors, became the rulers of Ontario's far north.

The Hudson's Bay Company, or HBC, owned northern Canada lock, stock and barrel of rum. Those near-royal factors governed their fiefdoms from forts, or factories. One of the first erected sat on an island at the mouth of the Moose River at the southern end of James Bay. Called Moose Factory, Ontario's first permanent settlement prospered beyond all imagination. Exchanging rum, guns, beads and blankets for high-quality winter-trapped furs with Native peoples turned out to be more lucrative than trading opium for tea in China.

New outposts were quickly constructed in the interior of Ontario's far north despite the man-eating bugs, bitter winter cold and sparse indigenous populations. At one of these interior outposts, an HBC factor noted in his log that Holland gin froze solid at –27°C, English Brandy at –32°C and West Indies rum at –35°C. Similar notations go a long way to explain why HBC shareholders rarely ventured "across the pond" to inspect their investment.

In the late 18th century, voyageurs from an upstart Montréal trading firm calling itself the North West Company (NWC) discovered they could paddle up the Ottawa River and push northeastward into the interior through 1600 kilometres of lakes and rivers. A fur trader's delight, save for the fact that the area was sparsely populated. Few Native peoples meant little or no trading, so the North West Company men did their

own trapping during the winter, when pelts are prime. Their system of navigable rivers and lakes meant in by September and out by June—with no bugs. A good plan, but one often soured by weather that even experienced voyageurs called weird.

To keep trappers working and furs moving east, the North West Company built outposts and lured southern Native peoples north. Not to be outdone and fearing encroachment on their domain, the Hudson's Bay Company pushed further into the interior, built more outposts and lured northern Native peoples south. In all, the two companies constructed almost 600 outposts, most of which became redundant after a few years of dirty tricks. Sending trappers to completely wipe out a competitor's stock-in-trade beaver population worked both ways and set the stage for a merging of the HBC and NWC in 1821.

Trapped-out beaver populations forced the fur business to move to Canada's northwest, probably a welcome event for trappers, because insects and bad weather had made northern Ontario a hell on Earth. Bugs and weather continued to plague those who remained until the middle of the 19th century, when northern Ontario's weather underwent a dramatic change for the better.

A CLIMATIC EXPLANATION

It's the Sun's Fault

Looking back, today's climatologists theorize a decline of solar sunspot activity to be the root cause of northern Ontario's extraordinarily foul weather that ended in the mid-19th century. This decline actually started in the 13th century, and the cold weather (at least in the Northern Hemisphere) that stretched from the mid-13th to the mid-19th century became known as the "Little Ice Age." European art galleries are full of paintings depicting skaters on the canals in Holland or the Thames River in London; paintings made during the peak of the Little Ice Age, when it was cold enough to freeze the rivers of Europe.

Scientists figure a decline in sunspot activity during this time prodded the ice-age-causing Quaternary Period into a last-gasp mini-episode of glaciation. Quaternary is the name given to the geologic period that runs from almost two million years ago to the present day. During that time, much of Canada was an ice rink, but the ice came and went. Temperatures dropped for a few centuries, and glaciers advanced; temperatures rose, and the glaciers retreated. This sequence repeated until the Quaternary relented around 10,000 years ago.

The Quaternary might have been down, but it wasn't out. According to climatologists, that one last Quaternary gasp we call the Little Ice Age continued on and off for almost 600 years. Climatologists have little knowledge about conditions in Northern Ontario before the voyageurs showed up in the 17th century, but between the cold and the bugs it must have been a most unpleasant place. Cold and rain lasted from spring to autumn, and if you failed to get out in time, you were toast, because winters were abominable.

The Year Without Summer

As the voyageurs trapped and suffered, the politicians got together in 1791 and finally put a name to the fertile strip of land that bordered the Great Lakes and reached a few hundred kilometres north to Rupert's Land, the name given to the big boggy bit owned by the HBC. They named the narrow strip Upper Canada, to distinguish it from everything east of the Ottawa River, which they called Lower Canada. The name Ontario, an Iroquois word for "beautiful lake" or "beautiful water" had been in use since 1641 and referred to the easternmost areas of the Great Lakes; the southern areas closest to the lakes were called Old Ontario. In 1840, our founding fathers passed an Act of Union uniting the colonies of Upper and Lower Canada into one central government to be known as Canada West and Canada East. So Ontario had three names: Ontario, Old Ontario and Canada West. Confused citizenry living in "Ontario" were much relieved when, in 1867, wise men gathered and passed an Act of Confederation that formed the Dominion of Canada and the province of Ontario.

Prior to Confederation and thanks to generally fine weather, Ontario, a.k.a. Canada West, had become an attractive destination for English and Scottish immigrants. Crops grew well, bugs were few and its citizens could walk around without fear of falling into a bog. Under generally sunny skies, the land prospered. Its citizenry grew cereal crops; cut oak, walnut and elm trees for export; and spent their earned pennies on English tea and treacle. But catastrophe, always just an Earth-burp away, struck in the early 19th century.

The disaster began on April 10, 1815, on the Indonesian island of Sumbawa, when the Tambora volcano exploded with a force unmatched in recorded history. *Ka-boom!* A 1.5-kilometre chunk of mountain vaporized and blasted 150 cubic kilometres of ash and debris into the upper atmosphere. Twelve hundred kilometres away, people heard its roar and worried that a storm might be coming. Indeed it was, because the biggest bang in recorded history killed an estimated 90,000 people, and the worrying had just begun.

What goes up comes down, but when volcanic ash goes far up into the stratosphere, that zone of atmosphere above the troposphere (the breathing layer), it actually orbits our planet, dims sunlight and drifts down slowly. The next year, 1816, darkening skies over the entire Northern Hemisphere signalled the start of the infamous Year Without Summer.

Snow fell in eastern Canada in June. Frost hit every month during the so-called summer, and drought struck in July and August. Crops failed, livestock died and residents of Ontario began to suffer famine. Without crops or livestock, small farms were abandoned, as Ontario farmers packed up and headed west, hoping for warmth and sunshine. With no farmers to buy their wares, local merchants resorted to cutting down the forests and selling the trees.

Before 1816, that part of Ontario south of a line from the Bruce Peninsula to Toronto was old-growth Carolinian Forest, an extension of the deciduous forest system that begins in the U.S. southern states. Oak and hickory varieties, black walnut, American chestnut, blue elm and some tropicals grew in abundance, and the local merchants, calling themselves lumber barons, took them all. Most southern Ontario towns and villages had lumber barons, and today their grand old mansions still line the main streets like memorials to hard times and greed.

Ontario's north country fared no better during the Year Without Summer, since all growing things require sunshine. No fresh buds 'n' bark meant hungry beavers, making them easy pickings for trappers. Estimates of 17th-century Ontario beaver populations run as high as 40 or 50 million, and during the latter part of the 18th century, a half million pelts were taken annually. During the Year Without Summer, trappers took almost all the remaining rodents, resulting in the end of the beaver hat craze. With the beaver suddenly a scarce commodity, northern trappers packed up and followed southern Ontario farmers west.

The Year Without Summer hit hard around the world. An estimated 200,000 people died in eastern and southern Europe from a combination of disease and hunger. China and India suffered famine as rice production fell. Ontario suffered, too, but it could have been worse.

No province-wide disasters, such as the Year Without Summer, have struck since. But localized weird weather persists, enough to feed a vivid imagination and provide a few "what if" thoughts. Could Ontario experience another Year Without Summer famine? You bet, and at any time, from a host of causes: volcanic eruption, comet impact, massive forest fires, immolation of North Korea from a nuclear accident....The list is long, and life but a gamble.

Ontario Weather Today

*Summer afternoon, summer afternoon;
to me those have always been the two
most beautiful words in the English language.*

–Henry James, author

ONTARIO WEATHER TODAY

WHERE DOES IT COME FROM?

Blame the Arctic

Ontario is high and slightly right of centre on the continental dartboard, and weather can arrive there from any point on the compass: cold from the north, warm and wet from the south, hot and wild from the southwest and dry as toast from the west. Sounds simple, and it is if the incoming systems miss one another. But if they collide, a war ensues—cold vs. warm—with Ontario supplying some extraneous influences.

The extraneous influence most affecting Ontario's weather is the Arctic Ocean, a vast and veritable desert in the winter months. The air we breathe has an affinity for water and makes room for it willingly—a variable amount, but it averages one percent of the entire atmosphere. Not much, but if it all fell out, Noah's flood would pale by comparison.

The frozen Arctic, like the Sahara and all deserts, has little or no surface water for air to absorb and is usually dry as dust, especially during the long arctic winters, when ice and little sunshine turn the high arctic air mass progressively colder. Cold air is only able to absorb diminished amounts of water, and if the seas are completely ice covered, the air gets nothing but the sweat from a seal's brow. Inversely, during warmer months, sunshine melts ice from large expanses of the Arctic Ocean, generating evaporation and increased humidity. The former months are freezing cold and dry as sand, the latter cold and damp—the very miserable continental and maritime polar air masses that can cause major weather problems for Ontario.

Continental polar air masses are cool and dry and form mostly in winter months over land. They are the principle instigators of Ontario's foul winter weather. They may also form during summer months, but they demonstrate a benign side by bringing fair skies to Ontario and the northern U.S. states.

Maritime polar air masses are both cold and moist because of their formation over the northern Pacific and Atlantic Oceans. Normally confined to the continent's northeast and northwest sections, maritime polar air is usually no threat to Ontario.

Ontario's concern is with the arctic and continental polar air masses. If cold arctic air pushes down from the Arctic Circle, it can force one of those polar air masses south. This also happens if the continental polar air mass becomes so large it simply spills out toward the south in what meteorologists call an arctic breakout. Not a serious problem; Ontario will have crappy weather, and its residents will need warm coats for a while. Problems arise only when crappy cold from the north meets hot and wet zipping up from the south or southwest. You've probably heard the saying: the sum is greater than its parts. That's pretty much the case when hot and zippy runs into cold, dry and crappy, because those parts can morph into monster low-pressure storm cells.

Low- and high-pressure cells entering Ontario from the southwest originate on the U.S. west coast and move northeastward over the Great Plains states. Those dry, cool highs provide beautiful sunny days, while the warm, wet lows are responsible for some of Ontario's weird weather.

Blame Alberta

Fast-moving low-pressure air sometimes enters Ontario from the west, off the prairies. Called Alberta Clippers because they originate in the lee of that province's Rocky Mountains, these fast-moving warm-air masses collide with cold air over the prairies and create volatile low-pressure systems with extreme cold fronts and high winds. Nasty, but luckily for Ontario, many Alberta Clippers either dissipate in the prairies or are roped by the jet stream and herded southeastward into the U.S. Great Plains states.

Albert Clippers can be tricky. They might head southeastward, meet some warm and wet air from the Gulf of Mexico and boomerang into Ontario from the south through the Ohio River Valley. Alberta Clippers can be bad-egg systems for Ontario, especially if they pass through the Great Lakes, where cold 'n' dry air is turned into cold 'n' wet air by the lake effect—cold winds move across warm lake water, picking up energy and water vapour. This can create the blizzard conditions common to the eastern and southern shores of those lakes. During the winter months, Alberta can spawn one or two clippers a week, and those reaching Ontario account for some weird winter storms: a quick drop in temperature, wind chill, snow and the awful sounds of shovelling and windshield scraping.

Luckily for Ontario, Alberta Clippers coming directly from the west are fast moving, with only the occasional lingerer. Unlucky (for Ontario) are the clippers that boomerang from the south, because they often set up house with northeastward-moving warm air masses and form either giant cyclonic storms or squall lines. In the cooler months, storm systems might pass through Ontario three or four times a week and cause high winds, heavy rain, tornadoes and massive snowfalls. Annual lake-effect snowfall from these storms can exceed 400 centimetres around Superior, Huron and Georgian Bay, and at times more than 100 centimetres can fall in 24 hours. During one event, December 7 to 9, 1977, a lake-effect snow squall, pushing high winds, dumped more than 100 centimetres on the city of London. A huge amount, but the blowing snow caused a doubling affect that closed the city for days.

Blame the Ohio River Valley

Alberta Clippers are a concern to Ontario primarily in winter; the concern in the warmer months comes mostly from the south and the Ohio River Valley. Storm systems have a thing for the Ohio River and gravitate to it like fish to water. During warmer months, the Caribbean and Gulf of Mexico become huge saunas as they're inundated by massive, westward-moving, low-pressure cyclonic systems that originate off the African coast. A few of these systems produce hurricanes, but most just belly in and push out existing weather.

Where does that pushed-out weather go? It goes looking for a watery path of least resistance. So it barrels up the Mississippi River to the Ohio River Valley and on into the Great Lakes. Warm, wet and feeling low, these weather systems roll up the valleys looking for a good fight with cold air and a long drink of water. In summer, they usually find that fight before they hit the Great Lakes, so they roll up thirsty and eager for stardom. These supercells are the celebrities of Ontario storm systems, the Big Kahunas, the property stompers that cost Ontario millions of dollars every season. Tornadoes, hail and floods all lie at the feet of the mighty supercell.

What is the difference between a supercell storm cloud and a run-of-the-mill thunderstorm? Presence. The supercell resembles the top of the mushroom cloud of a hydrogen bomb. All must bow to the supercell, and southwestern Ontario residents do a lot of bowing, because, outside of the U.S. Midwest, they get to see more supercell storms than anyone on the planet.

The supercell lives: it has a heart called a mesocyclone (a deep, continuously-rotating updraft), lungs called an updraft, a head bigger than some countries and legs that can stomp flat anything built by the hand of man. They are monsters—and residents of Ontario have front row seats to see them stomp. The next chapter will show you those seats like an usher in a theatre.

Ontario's Weird Weather Places

*Conversation about the weather is
the last refuge of the unimaginative.*

–Oscar Wilde

ONTARIO'S WEIRD WEATHER PLACES

THE NORTH

Northern Skies

The dictionary defines weird as involving or suggesting the supernatural; unearthly or uncanny—as in weird sounds and lights. Northern Ontario's weather can satisfy that definition almost anywhere at anytime. Although the south is more benign, it can be subject to periodic explosions of weird weather events because of its proximity to the Great Lakes.

Hop a plane north to Ontario's subarctic and you'll experience the supernatural, unearthly, uncanny and almost nightly displays of the aurora borealis—the northern lights. You've probably seen them from wherever you are, but those are mere shadows of the real thing. In the dark of a northern winter night, with no pollution nearby, the northern lights will knock your socks off. On a good night, the aurora will put on a show in colour and make you want to take up residence under these dancing lights.

The next morning you might see a sky full of diamonds, causing you to look for a job and somewhere to live. Diamonds in the sky, or diamond dust, are falling ice crystals—clear-air precipitation that appears mostly in pristine, pollution-free Arctic and Antarctic regions. In the right conditions, with no wind, the air can fill with sparkling jewels that seem to hang motionless—a fantastic, surreal experience. Move and they move; catch one on your tongue and taste infinity. Ontario's subarctic is about as supernatural a place as you'll find on this planet, and, hey, you can always find a job with De Beers, the diamond people. The company has a mine up there (in the James Bay lowlands near the coastal community of Attawapiskat) and are always looking for help. Diamonds in the sky, in the ground and almost every night a light show extravaganza…how surreal is that!

ONTARIO'S WEIRD WEATHER PLACES

Hudson Bay Frontier Region

From Haileybury, New Liskeard, Earlton and Englehart in the east to Hearst in the west and north to the shore of James Bay is Ontario's James Bay Frontier Region, an area paramount among Ontario's weird weather places. Sizzling summers, long, unimaginably cold winters (though resplendent with northern lights) and in between—all the weather in the proverbial thrown book.

The legendary El Dorados of Ontario are all here: the mines at Timmins, Cochrane, Kapuskasing, Hearst, Matheson, Cobalt, Kirkland Lake, Latchford and Iroquois Falls. Further north, across the great bog, is the mighty James Bay and the settlement where it all began: Moose Factory and its upstart neighbour Moosonee.

Moosonee was established in 1903 by the French fur and luxury goods company Revillon Frères to compete with the Hudson's Bay Company's Moose Factory, located across the Moose River. By 1909, the company had expanded to 48 trading posts, while the HBC was running 52. Competition heated up, and in 1936, HBC bought out the Revillon Frères company's fur-trading interests. A number of Inuit villages in Nunavut and northern Québec and Ontario are located on sites originally occupied by Revillon Frères trading posts.

If you like snow and atmospheric fireworks, Moosonee is your kind of place. Except for the months of June, July and August, it snows almost every day in Moosonee. A nice snow—dry, fluffy and perfect for snowmobiling out to the edge of the universe to be one with the stars. With no light or atmospheric pollution, you'll wander among millions of stars and stay up all night watching the dance of the veils, the northern lights. Be warned, though, that nice fluffy snow can sometimes last for days and really pile up.

ONTARIO'S WEIRD WEATHER PLACES

Peawanuck and Fort Severn are Ontario's most northerly settlements, and both are amazing "all about the weather" places. Native people recognize six seasons in the Peawanuck and Fort Severn area. Summer begins in early June, ends in late September and is followed by a short autumn. In October comes Mikiskow, the time of freezing. Winter begins mid-November, ends mid-March and can have wind-chill temperatures dipping to −80°C during those wicked months of January and February. Spring is March to April, followed by Mennoskumin, the time of thawing, when the tundra once again becomes habitable.

The population of both settlements is around 500, mostly First Nation Cree and Ojibwa—delightful people with sunny dispositions, in spite of having to live on the edge of the known universe. Just kidding, but that's how it seems on those clear, cold, zillion-stars-in-the-heaven nights while you watch the aurora borealis dance overhead in living colour. You stand there speechless, feeling small and insignificant, but with your faith in the Creator renewed. What a place, and if not for the teeth-snapping −40°C, you might want to stay. Come morning, the air, warmed by sunshine, fills with sparkling diamond dust and your thoughts turn to looking around for a house.

Aside from that house, you'll need a vehicle to get around, and though there are no roads up there, there exists an amazing network of wintertime icy highways—3000 kilometres of roads over frozen water. More amazing is the fact that you can actually drive to Fort Severn and Hudson Bay from wherever you are. Only in winter, mind you, and it's a gruelling trip. Start in Manitoba's capital city of Winnipeg, drive north, skirting massive Lake Winnipeg, and head toward Thompson, Manitoba. Here, you'd better overnight, warm up and prepare for the remainder of the journey because the track north is a gravel road. A few hours travel up that road, hang a right (go east) onto the ice road to Fort Severn. Only 500 kilometres to Fort Severn, and, with any luck, you'll find the road free of trucks and make it there in 10 hours.

Weird stuff, ice, because even metres thick it's still a liquid and will behave like one if subjected to stress. Overweight trucks going too fast will sometimes create a wave in the ice road, push it ahead and pile it up until the ice fractures. This happens occasionally to impatient truckers, but you'll be all right—your car or van is not heavy enough to make a wave. If you dare the ice road, take extra fuel and emergency supplies, because sometimes the weather can turn. One minute, the sun is shining and you're worried about snow blindness; the next, you can no longer see the road for snow. On such occasions, hunker down and wait for the plow that will come along, eventually, once the snow has stopped. Hey, you wanted adventure, and there it is—an endless road across an endless world of snow and ice.

ONTARIO'S WEIRD WEATHER PLACES

THE GREAT LAKES

In the Beginning

You don't have to travel to Ontario's north to experience weird weather, because there's plenty in the south. Ontario is huge and contains about 35 percent of the world's fresh water; of that amount, two-thirds flows through the Great Lakes—sometimes fast, sometimes slow, but always heading for the sea. The Great Lakes are the youngest large, natural feature on the North American continent. They owe their existence to three distinct periods of advance and retreat of the early Wisconsin Glacier, an ice sheet that was two to three kilometres thick.

The Wisconsin Glacier began east of Hudson Bay in Québec and Labrador approximately 65,000 years ago. It advanced across North America, where it remained for a period of roughly 15,000 years before retreating. About 40,000 years ago, the Middle Wisconsin Glacier advanced to cover half of North America; it hung around for 8000 to 10,000 years before retreating. Finally, the Late Wisconsin Glacier advanced some 18,000 to 20,000 years ago and hung around for 10,000 years.

The various Wisconsin Glaciers gouged the Great Lakes to bedrock, except for Lake Ontario, which already had a riverbed depression. In addition, the sheer weight of the ice caused the rock to depress and tilt back toward the glacier. Off and on during these periods, glacial meltwaters covered a huge area of North America. Over time, they drained away, leaving four glacial Great Lakes depressions (plus Lake Ontario) and various drainage routes, including the St. Lawrence River. During the last retreat, glacial water ran to the sea until sea levels fell, at which point water began accumulating in the depressions and drainage resumed through the original watershed—the St. Lawrence River.

In the case of glacier-buried land, what goes down comes up again, and the lakes are rebounding at a rate of roughly 50 centimetres every century. Not a problem, but it does have an effect on lake levels because of the tilt factor. The lakes are rebounding faster on the north side, causing water levels to rise on the south, which is resulting in losses to American property owners. Okay, so this isn't exactly weather information you can use today, but it's nice to know you can bequeath your Great Lakes lakeshore property (on the Canadian side) to your great, great grandchildren without worrying that they'll be flooded out over the long term.

Lake Superior

The legend lives on from the Chippewa on down of the big lake they called Gitche Gumee.

The lake, it is said, never gives up her dead when the skies of November turn gloomy.

–Gordon Lightfoot,
"The Wreck of the *Edmund Fitzgerald*"

Lake Superior, at the head of the Great Lakes, is 563 kilometres long, 257 kilometres wide and, with an average depth of 147 metres, is more inland sea than lake. Superior is icy cold in all seasons with a resident water time—the time water spends in a lake—of 100 years. Cold water does not evaporate quickly and is generally weather benign. This is Superior during the summer months, when its water is slowly heated a few degrees by sunshine. Still icy cold but warmed enough to stay mostly ice free during the winter months, Superior's open waters interact with warm, wet-weather cells moving north from the Gulf of Mexico, turning them into monster ship-sinkers.

Obviously that's a problem for mariners but hardly weird—except during the month of November, when two weather systems converge on Lake Superior like clockwork: a southeast blaster from Alberta and a storm system originating east of the Canadian Rocky Mountains. This combined weather system then invariably joins warm southern air to form the infamous November gales on Superior, which can generate winds topping 200 kilometres per hour and waves exceeding 15 metres high. Monster storms you can almost set your watch by—but make sure you're on dry land when you do the setting.

Superior's icy waters move ever eastward over an increasingly shallow lake bottom. In summer months, its water is warmed by sunshine, which can generate massive lake-effect rain. Warm water heats the dry westerly winds coming off the prairies, and the warmed air absorbs evaporated water vapour that condenses when cooled as it rises over land. In winter, cold air moving over the slightly warmer waters of Superior also absorbs water vapour, which freezes and dumps as snow on land downwind.

That's lake effect, and if cross-country skiing is your thing, Superior's north shore in winter is your place, because it can receive more than 900 centimetres of the white stuff. Inversely, summers on Lake Superior are wonderful, with clear water, great scenery and only the odd weather event, such as pea-soup fog. Fog is the norm on Lake Superior as a result of warm air in contact with icy water; it's not considered weird weather.

Weird happens during the cold seasons, when monster storms and high waves have you thinking North Atlantic. Some storms are so severe they create a storm surge, which can be catastrophic to shorelines. Storm surges are not confined to Canada's coastlines. They can occur in large bodies of fresh water, especially in the Great Lakes, where they're called *seiche* (pronounced "saysh"). Think water sloshing in a bathtub with a low-pressure storm at one end instead of a kid. All the Great Lakes have a perpetual seiche of a few centimetres that goes mostly unnoticed, but those annual Superior storms can tilt lake water more than a metre. This weird effect can leave boats temporarily hanging from their docks, chew up lakefront property and sweep dog walkers, fishermen and sun seekers off beaches.

Lake Huron

Moving eastward toward the sea, Lake Superior's icy water funnels through the St. Mary's River into Lakes Huron and Michigan. This channel is a busy spot; every year more than 120,000 ships move back and forth through locks between Huron and Superior. But the water is in no hurry. It rolls on gently into two inseparable lakes that resemble saddlebags laid over the upper portion of the state of Michigan. Lakes Michigan and Huron are hydrologically one lake with two names, connected by the 90-metre-deep Straits of Mackinac. The pair are shallower than Superior with an average depth of 59 metres (Huron) and 85 metres (Michigan). Shallower means warmer water to feed storm cells, and both Huron and Michigan see their share of monsters.

On November 8, 1913, an arctic cold front collided with a rare low-pressure anticyclonic (clockwise-flowing) air mass moving in from Lake Erie. The next morning, it charged forth as a monster cyclonic storm (air flowing counter-clockwise) and moved toward Lake Michigan. Area newspapers got the alert the day before, but wrote it up as…expect intermittent rain.

Big surprise for everyone when Frankenstein's monster turned into what meteorologists call a "weather bomb" (when atmospheric pressure falls rapidly, which often occurs during a cyclone or hurricane) and suddenly leaped from the horizon, blasting out 150 kilometre-per-hour winds and blinding snow. Weather historians call it "The Great Lakes White Hurricane," while to old-time sailors it was "Black Sunday." Although not a true hurricane, it did live up to the name by lasting more than three days and doing damage befitting a category cyclone.

"Great White" devastated the shorelines of Lakes Michigan and Huron. Reaching Lake Erie and even warmer water, Great White hung around for two days, dumping snow metres deep onto London and completely burying Cleveland, Ohio. Done there, it headed across Lake Ontario and up the St. Lawrence River, where it ran out of steam. In all, the Great Lakes White Hurricane sank as many as 40 ships, drowned some 250 people and caused millions of dollars in damage, a tidy sum in those bad old days. Cleveland was paralyzed and without power for days.

Cleveland lay in white and mighty solitude, mute and deaf to the outside world, a city of lonesome snowiness, storm-swept from end to end, when the violence of the two-day blizzard lessened late yesterday afternoon.

–*Cleveland Plain Dealer*, November 11, 1913

In 1996, a rare hurricane developed over Lake Huron. Dubbed "Hurricane Huron," it formed September 11 in Lake Superior as a weak cyclonic storm with an icy core. Two days later, it had moved on to Lake Huron. Cloud heights increased, wind speed picked up and the storm's surface cold front became occluded (stationary) and reached all the way to Pennsylvania. By the afternoon of September 14, Hurricane Huron had all the attributes of a tropical cyclone: an eye 30 kilometres wide and a connecting spiral of clouds forming a wall.

The storm surge from Hurricane Huron caused extensive shoreline damage to the Great Lakes, and massive rain—more than 100 millimetres in some places—and flooding hit surrounding areas, including parts of Ontario still recovering from the effects of a September 8 overrun by the dregs of Hurricane Fran. A double whammy, but Hurricane Fran's muscle might have kept the Huron cyclone mostly stationary, thereby saving Ontario from a far worse pummelling. The tail end of a Caribbean hurricane like Fran is one thing, but a homegrown cyclone crashing through the back door could have been catastrophic.

Both Lakes Michigan and Huron are susceptible to seiches, and while no seiche fatalities have been reported for Lake Huron, its sister lake, Michigan, has experienced more than a dozen. Seiches have plucked bathers off beaches, and on June 26, 1954, a three-metre-high seiche rose out of calm water and washed away eight fishermen from a Chicago breakwater, drowning the lot.

Georgian Bay

Half in the south country, half in the north, Georgian Bay abounds in weird weather: tornadic events, downdrafts, water spouts, supercells, lake-effect blizzards and massive storm fronts that, over the years, have sunk hundreds of ships. This shallow bay is a huge part of Lake Huron—almost as large as Lake Ontario—and a straight reach from Lake Superior from whence it receives an almost constant westerly wind. Georgian Bay is all about that west wind—bent pines, rocks eroded into weird formations and normally placid water whipped into ocean-sized waves in minutes.

Shallow with constant westerly winds, Georgian Bay was a curse to lake mariners of old but a boon to trappers. This was the southern route to the north selected by the Hudson's Bay Company to access those 1600 kilometres of navigable lakes and rivers discovered by the North West Company. Paddling was arduous, though the 30,000 islands scattered throughout the Bay provided shelter from wind and storms. In by September, out by June—that was the order of the day for the fur traders. Any delay might have those voyageur trappers in the clutches of an early fall gale and up to their necks in drifting snow. Those were dangerous times, and weird weather could arrive from Lake Superior during any season.

Times change, and so does the weather; it has warmed and mellowed. But if you like early-season cross-country skiing, or snow is your thing, Georgian Bay is your spot, because there is an excellent chance those westerly gales can put you up to your neck in the white stuff. Storm cells rolling off icy Lake Superior onto warmer Georgian Bay will sponge up massive amounts of water and dump it as snow, rain or hail (season dependant) as they reach cooler ground. Wind alone will accomplish this lake effect most any time, but storm cells pick up huge amounts of water and can dump everything but the kitchen sink, creating deluge or blizzard conditions.

Inversely, the summers on Georgian Bay are a tourist's delight: warm, sparkling water, 30,000 tiny islands to explore and sunshine galore. Two fingers of land that delineate the Bay from Lake Huron deflect the westerlies blowing off Lake Superior:

the Bruce Peninsula and a paradise called Manitoulin—one very large island.

Manitoulin Island

Manitoulin is the largest freshwater island in the world (2765 square kilometres) and a place with more island lakes than anywhere on the planet (110)—including Lake Manitou, the largest lake found on a freshwater island. Manitoulin Island is another all-about-the-weather place, and nowhere on the planet is this more obvious. Rock everywhere, all 4.5 billion years old, scraped by glaciers, cracked by ice and weathered by constant westerly winds and storms. To be there is to be standing in a Group of Seven painting of weathered rock and wind-whipped pines.

During the horse-and-buggy days, the great forests of Manitoulin Island supplied Ontario and many U.S. states with lumber, shingles, hop poles and road pavers. Hop poles are two-metre cedar poles used for training hop vines, hops being a vital component of beer. Road pavers were cedar poles cut into one-foot lengths, set upright into a street and backfilled with dirt. Cheaper than brick, wooden road pavers were used by many Ontario and U.S. communities until the

advent of the motorcar. Automobiles needed fuel, and the heat cracking of oil into gasoline generated a waste product—tar—that became widely available for road topping. Manitoulin sawyers shipped boatloads of poles and pavers to many points from Little Current, the principle island port for schooners and steamships.

The number of shipwrecks littering the bottoms of the Great Lakes is somewhere in the area of 5000—a staggering quantity of ships. The numbers of lives lost is unknown and probably equally staggering. The Great Lakes are a vast graveyard, and each wreck is a grisly testimony to the power of storms. Nowhere is that testimony more profound then in Georgian Bay. Of those 5000 wrecks, a good many lie on the Bay's sandy bottom amid the vast minefield of tiny islands.

At 4:00 AM on November 22, 1879, the *Waubuno*, a 36-metre side-wheel steamer, put out from Collingwood (headed for Parry Sound) with 24 passengers and crew. She encountered a storm of such ferocity that waves ripped off the ship's superstructure. Days later, when search boats located some of the wreckage of the *Waubuno*, they found only the hull and no bodies. A few years later, on September 14, 1882, the *Waubuno*'s replacement vessel, a larger steamer named *Asia*, put out from Collingwood for Sault Ste. Marie with 125 passengers and crew aboard and ran into a similar storm. The *Asia* foundered and capsized north of Parry Sound; only two survived. These are only two of the many steamers that sailed into Georgian Bay, encountered weird weather and were never seen again until found by modern-day divers.

Wasaga Beach

Located at the bottom end of Georgian Bay is Wasaga Beach, one of the world's great beaches, and the residents of Ontario have the weather to thank for its magnificence. Wide, white and hard-packed enough to support vehicles, it stretches some 14 kilometres and is the longest freshwater beach on the planet. The water is great—clear, refreshing, sandy-bottomed and ankle deep for almost a half kilometre. Shallow, too, which explains the great beach.

The constant westerly winds create a smooth bottom current across Lake Huron that becomes erratic when encountering those thousands of tiny Georgian Bay islands. Called channelling and tunnelling, this undulating current picks up bottom sand and throws it off as the current sweeps around the bottom end of the bay. Placer gold miners call that a "throw"; the water turns, while heavy particles—sand or gold—in the current keep on trucking ahead to become either beach or bonanza.

Lake Erie

When water from Lake Superior finally reaches extremely shallow Lake Erie via Lake Huron and then tiny Lake St. Clair (and the rivers St. Clair and Detroit), it cooks in the sunshine. Lake Erie has an average depth of a mere 19 metres and a resident water time of only two to three years. Warm, shallow and long, Lake Erie is a weather maker and a main crossing point for northeastward-moving weather systems.

Storm cells are transient beasts. They like to move fast, and Erie gets them from three directions: warm and wet from the Caribbean up through the Ohio River Valley; warm and wet (or dry) from the southwest across the U.S. Great Plains; and dry and icy cold from the Arctic. Cold or warm matters little, because all transient air loves the warm waters of Lake Erie, and each storm cell, or combination, will create foul weather that often becomes extreme and weird.

Because Lake Erie is a long, shallow lake, it's prone to wind-driven seiche that can sometimes exceed five metres. During a November 1972 storm, northeast winds reached a constant speed in excess of 50 kilometres per hour, and the seiche inflicted millions of dollars damage to both U.S. and Canadian shorelines. In September 1985, a wind-driven seiche overran Long Point and swept 40 cottages into the lake. Luckily, it was off-season and only property was lost. But if you want to experience the extremely weird, hop a ferry to the middle of Lake Erie, to a 10,000-acre island of anticipation called Pelee.

Pelee Island

Pelee Island in Lake Erie is the southernmost populated point in Canada. At 41° north latitude, this large island is 775 kilometres south of the latitude of Victoria, BC, and at the same latitude as northern California. Warm, humid and handy to Ohio River Valley storm systems, Pelee can be a cauldron of atmospheric activity. This makes it ideal for observing supercell thunderstorms and the weirdest of small tornadoes, the elusive waterspout.

I have been to Pelee Island seven times, seen water twisters on three occasions and have always left the island thinking they ought to be promoted as a tourist attraction. During a week in late September 2003, dozens of waterspouts appeared in the Pelee area, and folks there saw a never-to-be-forgotten show.

Waterspouts are an atmospheric treat. Powerful, snow white and writhing like conga dancers, they put on a fantastic and relatively safe show because they seldom come ashore—just don't sail out to meet one. Waterspouts are tornadoes over water caused by the same wind shear that creates their more conventional big brothers. Usually not as powerful as tornadoes, they're more prolific because of the wide-open space of the lake and a convenient supply of thermal energy. Lake Erie pumps a lot of heat into the air, especially during the summer months—the best time to observe these wondrous waterspouts.

If you do go looking, keep in mind that their winds are dangerous, but hey, they hardly ever hit land. Not so with their big brothers, and although Pelee Island has experienced a few tornadic overruns, there has been nothing serious—so far.

Pelee Island is Ontario's southernmost contribution to that deciduous tree belt called the Carolinian forest and enjoys more frost-free days and warmer winters than anywhere in Ontario. Folks call this stretch of the northern shore of Lake Erie "the Banana Belt." It comprises an area stretching from Windsor to London and from the lakeshore to a few kilometres inland. The area is good for plants, great for storms, and in the colder months, a fantasyland of snow squalls.

Farmers on Pelee Island grow soybeans, a low-to-the-ground crop able to withstand inclement weather. They grow wine grapes as well, on squat vines that are also good at weathering storms. The attendant winery produces a surprisingly good vintage, and all the restaurants do a decent picnic basket. Good beaches, great wine, food, warm water for swimming, lots of sunshine (when not storming), the occasional waterspout for entertainment and a stand of cacti indigenous to Ontario. What a fabulous spot!

Pelee Island is so nice you might want to stay a week, which might be required to see weird weather. Do stay, because Pelee Island is all about weather. Its 300 permanent residents talk about it constantly, but one word is always whispered with reverence…tornado. Locals are always anticipating the arrival of a big twister and with good reason, as the island is front 'n' centre of a tornadic stomping ground.

Ontario has four tornado alleys: three minor ones—Lake Superior to Algoma, Lake Ontario to Parry Sound, Lake Ontario to Renfrew County—and one bowling-for-dollars Big Kahuna—the Lake Erie to Georgian Bay alley. If storm chasing is your thing, this is your spot. It begins in Essex County (the southwestern-most segment of Ontario), runs northeastward past Lake St. Claire and hugs the shore of Lake Huron all the way to Simcoe County on the southeastern shore of Georgian Bay. Seek the beast, and ye shall find it in this region. But your chances are better at either end, because Essex and Simcoe Counties have seen the most recorded twisters. Tornadoes are moving atom bombs of destruction and about the scariest weather phenomenon on our planet; if you go out looking, be careful.

Point Pelee National Park

On the Canadian mainland slightly east of north of Pelee Island is a skinny, 10-kilometre-long triangle of sand. It's Canada's smallest national park, lies on the northern shore of Lake Erie and contains the southernmost bit of mainland Canada. Point Pelee's 20 square kilometres feature broad marshes, savannah grasslands and a dark Carolinian forest that will have you searching the trees for monkeys. No monkeys, but rare plants, animals, reptiles and birds abound. It's a jewel of a place, almost lost to overexploitation.

During the 19th century, local farmers put orchards and crops into loamy areas and melons into the sandy parts and overran Point Pelee's ecological variety. Overtilling of the sandy soil denuded whole areas of vegetation, and, when discovered by birdwatchers at the turn of the century, Point Pelee was mostly sand dunes with a narrow section of woodlands overrun by livestock. Recognizing that the area was an important flyway for both migratory birds and the monarch butterfly, the birdwatchers fenced off the woodlands and petitioned Ottawa to proclaim the area a national park. Lucky for Ontario, those birdwatchers had friends in high places, and in 1918, Ottawa passed an act preserving the area for ecological reasons.

So Point Pelee got some breathing room and began to rehabilitate itself, but in the mid-20th century, tourists discovered Pelee and the degradation went into overdrive as three-quarters of a million people trooped into the park every week, and cottages sprouted like mushrooms. Ottawa came to its senses in 1970 and began to rein in the commercial, tourist and weekender traffic. Since then, the federal and provincial governments have removed six fish-processing plants, 20 kilometres of roads, 400 buildings plus orchards and farmlands, and have made a concentrated effort to properly manage the park. Point Pelee is a shining example of wilderness rejuvenation; an icon of what can happen when people stop taking and start giving back.

Point Pelee's marshland is the largest wetland area in Canada and an international tourist attraction that now hosts a more sensible 300,000 annual visitors. Visitors can rent canoes or walk the marsh on a specially constructed raised pathway, but entry into ecologically sensitive areas is prohibited. Point Pelee lies within a zone known as the humid continental climate, so hiking the boardwalk or canoeing will bring on a sweat. But humidity is what makes it a weird-weather place, as evidenced by Pelee's forests, grasslands and beaches.

The Point Pelee peninsula contains dry woodland, savannah and swamp cut through with waterways; a magical place that's home to more than 700 species of plants, 90 birds, rare amphibians and mammals. Inland, the swampy forest gives way to a Red Cedar savannah that features rare grasses and the prickly pear cactus (also found on Pelee Island). After the savannah comes the dry Carolinian forest that gives way to a more familiar mixed wood of sugar maples, white pine and basswood trees. Point Pelee is a unique Ontario microclimate (remember, it's part of the "Banana Belt") that owes its very existence to weird and nasty weather.

The beach tells the story of why Pelee is such a wondrous place. It's a cuspate sand spit (a bit of land having cusps or points) that juts into Lake Erie like a hitchhiker's thumb and was formed by either glaciation or convergent lake currents. Point Pelee's 20-plus kilometres of beaches are under almost constant attack by storms, and today's beach can be gone tomorrow—only to be back the next day full of driftwood and bits of shipwreck flotsam.

In the days of sailing ships, Pelee Passage, the shoal-dotted stretch of water between Point Pelee and Pelee Island, was considered by sailors to be the most treacherous water on Lake Erie. More than 275 ships litter the bottom, and your chances of finding their bits on the beach are fair to good. But be warned, any good bits and bones must be turned over to park officials or you could find yourself watching Pelee's big storms from the Big House.

Point Pelee is front and centre at the northern end of the Ohio River Valley system and a great place to observe weird weather, because almost every other day during the summer months you can see supercells form over Lake Erie. Sometimes they dissipate, but often they grow into monsters that deserve an audience. Go see them and be better for the experience.

Rondeau Provincial Park

If you want to see southern Ontario through the eyes of its early Native peoples and the voyageurs, this is your spot. Rondeau Peninsula, some 50 kilometres up the shore of Erie from Point Pelee, extends almost eight kilometres into Lake Erie and encloses a bay of significant ecological importance. Founded in 1894, Rondeau is Ontario's second oldest provincial park (after Algonquin) and is the province's largest remaining tract of what used to be the deciduous forest region of Canada—the original southern Ontario. Rondeau is a place created and sustained by weird weather. It's another cuspate spit, composed of sand and gravel pushed there by converging winds and water currents. It's alive—always moving a little here, a little there, and changing shape.

Rondeau Park edges Lake Erie with undulating sand dunes formed by the lake's rise and fall. Dunes give way to soggy dips and dry ridges, marshlands, the inner bay, black oak savannahs and the Carolinian forest. Rondeau's forests and marshlands are home to rare plants, birds, animals and reptiles found nowhere else in Canada. Rondeau is also on the migratory flyway and a birdwatcher's paradise. Bring your spotting card to Rondeau and you'll have opportunity to check off more than 250 species of birds—some native but most migratory. Aside from wildlife and rare species, Rondeau is also front 'n' centre at the northern end of the Ohio River Valley system, and, like Point Pelee, is a great place to observe weather getting weird.

Sand Hill Park

If you continue northeast along Lake Erie's shoreline, you'll eventually run into (or see directional signs for) Port Burwell on regional road 42, the lake road. Head into town, but keep going about 12 kilometres east along the shore. Why? Because you just have to see what Lake Erie's weird weather has created for your day in the sun—Sand Hill Park, a.k.a. the Houghton Sand Hills. Kids and dogs go crazy over this place, and maybe you will too, if you're not too old in spirit.

As the sign says, it's a sand hill, but from a kilometre away you'll already know those words are understating the situation, because you'll be able to see it over the treetops. The thing is enormous, a pile of sand 120 metres high and about four kilometres long. Weird, because it's not part of any dune or beach system—just a giant sand pile on a beach surrounded by farmland.

How did it get here? For the answer, remember the wind-turbine farm you passed a few miles back. That's right, it's the wind. You might recall (from the wind primer earlier in this book) that air rushes from high pressure to low—and that's what is happening offshore. The water is so shallow and loaded with sandbars that it heats up faster than the surrounding deep water, which in turn heats the air causing it to rise quickly. Fast-rising air draws in fast-moving air that picks up tiny grains of sand from wave tops and drops them onto the shore.

Lake Erie contains a lot of fine sand scraped from rocks and bulldozed into place by glaciers. Sand moves around the lake bottom at the whim of currents, and sometimes these currents converge and will form cuspate spits if enough piles up to break the surface. Houghton would be a cuspate spit if the currents were a bit stronger; as it is, the spit is actually a half-metre under water. It makes for interesting swimming, because you can walk out a long way in only knee-deep water. Careful though, because storms have a way of shifting sand to form deeper channels called bars.

Take along the kids and dog and enjoy a day in the sand pile. However, you had better warn them to stay away from the sulphur springs found here and there along the shore, or you might not want them back in the car.

Long Point

I grew up near Long Point (20-plus kilometres east of Sand Hill Park). It was my summer playground for years. My dad liked fishing, and for that he used a 24-foot cabin cruiser with a flat bottom. The perfect craft for exploring the swampy waterways of Long Point, a 40-kilometre spit of sand sticking out into Lake Erie like a bent thumb. Inside the bend is Long Point Bay, a stopping-off place for half the world's migrating duck population and the best small-mouth-bass fishing on the planet. A great place for ducks, fish and learning to swim the hard way.

I was six years old, out for an afternoon of fishing with my dad and his buddies, when for no particular reason my father chucked me off the back of the boat. Talk about panic—I was sure my dad had lost his marbles, and all I could see of land was a thin line on the horizon. I remember thrashing about and actually swimming a bit before my dad yelled for me to stand up. Hey, I was only six—how was I to know the boat was floating over a sand bar in less than a metre of water.

Long Point is another hot spot for weather—lots of storms and strong bottom currents for moving the sand that forms

the spit and its marvellous inner bay. In the bad old days, the inner bay was a major destination for sailing ships seeking refuge from Erie's many storms. Refuge from storms meant finding the Old Cut, a storm-created channel through the Point and into the bay. Those that failed to find it litter the bottom around Long Point, and sometimes their bits and pieces wash ashore like driftwood. If you want summer fun, try beachcombing the Point's beaches after a storm has scoured the sandy bottom. You might even grab a metal detector and try your hand at treasure hunting. It's said that, in the late 1700s, a fur trader named David Ramsey—fearing a (justified) attack by angry Native peoples—buried a chest of gold in the sands of Long Point and never returned for it.

During the late 17th and early 18th centuries, the area encompassing Long Point (now part of Norfolk County) was called Long Point Settlement and home to trappers, fisherman and loggers. When these area settlers moved inland to farm, Long Point was all but forgotten until 1866, when the province sold its 17,000 acres of mostly marshland to a group of sportsmen who founded the Long Point Company and created a private duck-shooting preserve. This turned out to be a fortuitous undertaking, because it saved the Point from loggers and the ruination that befell surrounding areas. In 1979, the Long Point Company, in a democratic moment, gave half its holdings to the Canadian Wildlife Service on condition that the site be preserved in its natural state. Now anyone can hunt ducks in season, and blind rental fees can be found online—quack, quack.

Aside from ducks, Long Point is all about weird weather. Strong converging lake currents push sand around like rice in a soup, creating huge expanses of shallow flats and bars. Long Point, like Point Pelee and Rondeau, is another cuspate spit formed during the last 12,000 years that constantly bends to the whim of storms. Unlike Rondeau and Pelee, what changes most at Long Point is the surrounding bottom. Mostly shallow, but often cut through by storms, the bottom sands of Long Point reach out into Lake Erie like a moving hand that, in the old days, found and sank shipping schooners on a regular basis.

Between 1850 and 1885, no less than 50 ships ran afoul of Long Point's shoals and bars with great loss of life. The original *Edmund Fitzgerald*, the namesake of the modern-day freighter made famous by singer Gordon Lightfoot, sank in the Old Cut in 1883 with the loss of all hands. In 1906, a huge storm sealed up the Old Cut, but by then ship traffic on the lake had dwindled, and by 1916 the Old Cut lighthouse, built in 1879, was decommissioned. These days, the shallows are well marked on charts and, though the odd luxury cabin cruiser comes a cropper, accidents are rare and the shallows are more an ecological benefit than hazard to navigation.

Water flowing across the shallow flats and bars into Long Point Bay is super-heated by sunshine, which imparts an almost tropical environment to the area. Some areas of the spit and its surrounding marshlands will have you thinking Florida Everglades, as will some of its critters. Long Point is a World

Biosphere Reserve and home to 80 species of birds, 60 species of fish, rare snakes, plants and amphibians. Monarch butterflies use the spit as a resting place on their momentous journey to Mexico; 50-kilogram catfish and dwarf deer are common.

During the early days of the 20th century, when the travelling circus visited almost every town in Ontario, an odious feature of the sideshow was the snake pit. Long Point was where they got their slithering specimens by the thousands: black, king, rat and the odd rattlesnake. Long Point is an ecological oddity; a place shaped by violent storms, swift lake currents and a refuge for eight million migratory birds.

ODD LANDFORMS

The Niagara Escarpment

Some 400 million years ago, a great sea called the Michigan Basin covered an area from the western edge of Lake Michigan to the northern shores of Lake Huron and down through Ontario into New York state. During the Cretaceous Period (100 million years ago), most of the sea drained away and left behind a raised shoreline. Capped by hard dolomite limestone, the shoreline's underlying soft lime and sandstone eroded until it formed an escarpment or cliff structure called a *cuesta*.

Ontario's portion of this cuesta—the Niagara Escarpment—runs from the Niagara Falls area near Lewiston, New York, all the way to Manitoulin Island and boasts a phenomenal 720-kilometre-long hiking experience called the Bruce Trail—Canada's oldest and longest trail. The escarpment and path is a UNESCO World Biosphere Reserve and is well marked from its start near Queenston to its end in the village of Tobermory at the tip of the Bruce Peninsula. Some of the

wildest and best scenery in southern Ontario can be found along the Niagara Escarpment trail: 60 waterfalls, untouched forests, wildlife, ethereal solitude and great vistas of Ontario's weird-weather creations.

The actions of weird weather can be seen at various spots along the trail, starting with a section of southern deciduous forest. Here, one can relive Ontario the way it used to be, before settlers arrived with axe and saw.

From the heights you can look down onto land bordering Lake Ontario, see kilometres of vineyards and think of France. Indeed, the vines are French and survive because of a microclimate—Lake Ontario and the escarpment serving to modify the climate into a bit of France in the midst of a lot of ice and snow. During the fall, air warmed by the lake rises up the escarpment, pulling more warm air onto the vineyards and lengthening the growing time. During spring, the same effect pulls cool air off the lake, delaying the arrival of buds and protecting them from frost damage.

North of the Niagara Peninsula, in the escarpment's midsection, the microclimate dwindles, and the area known as the Great Lakes-St. Lawrence Forest Region becomes populated by birch, aspen and white cedar. During the early 1800s, this was a prosperous region, until the inhabitants cut so much forest the area suffered repeated flooding—a cautionary tale, to be sure.

About halfway along the trail is Dufferin County. This area is the highest plateau in the region and the source of five of southern Ontario's rivers: the Humber, Grand, Credit, Nottawasaga and Saugeen. Heading north, hikers pass through Mono Centre and onto the wondrous Mono Cliffs, a place where Ontario's weather has hewed some fantastic geological features: crevice caves, waterfalls, deep, fern-filled grottos and a complete separation of the escarpment by erosion over thousands of years. At Nottawasaga Bay, hikers reach the escarpment's highest point of 350 metres.

Past Owen Sound, the escarpment follows the shore of Georgian Bay, and the forest turns boreal, with jack pine blending into great stands of balsam fir and white cedar. The Bruce Peninsula is considered to be the most beautiful stretch of the trail—the green waters of Georgian Bay contrast with the escarpment's dolomite cliffs. Close to Tobermory, some white cedars growing from rock crevices and cliff sides are more than 1000 years old. These must be treated with great respect and left alone.

Everything looks different on the Bruce Peninsula because the weather is different here from other areas of the trail. Up here, the land suffers the westerly winds from Lake Superior, and there is evidence of wind and storm erosion everywhere. Weathered cliffs, rock pinnacles and deep crevices all bear testimony to the power of those westerlies. A power you might experience if you're on the trail for the long haul.

Cheltenham Badlands

Here is your opportunity to visit Mars without leaving the comfort of your own planet. Simply drive an hour northwest of Toronto, then along Old Baseline Road north of the village of Cheltenham, and there it is—Mars, the red planet! Okay, so maybe it looks more like the Alberta badlands painted red, but it's an odd sight. Red mounds, greenish gullies, shades of grey and the odd stunted tree make the Cheltenham Badlands one of Ontario's weirdest weather places. Stunned visitors always ask: how did it get here? Not you, because I'm going to explain its unfortunate creation by rain, wind and the hand of man.

The paint-box red of the Cheltenham badlands is iron oxide scraped from rock by glaciers and deposited onto the floor of a great prehistoric sea as the glaciers retreated some 430 million years ago (during the Ordovician geologic period). As lots of time passed, that iron-oxide bottom soil became compressed into a type of rock called Queenston Shale, a major component of the nearby Niagara Escarpment. Ordovician shale underlies most of south-central Ontario, but only near the escarpment does it lie so close to the surface.

Pull off the surface vegetation, scrape the ground with your foot, and presto—red shale. (The narrow greenish bands you see here and there reveal where groundwater has altered the rock's chemistry from red to green iron oxide.) Chop down lots of trees, pull away too much vegetation, and eventually weather will erode an entire area into a Martian landscape, which is what happened north of Cheltenham at the start of the 20th century. Poor farming practices—too many sheep probably, but it could have been cattle—caused overgrazing, and weather erosion did the rest. What you see at the Cheltenham Badlands took less than a century to accomplish and is primarily due to the hand of man. The prehistoric quality of the place is nothing but a mirage.

EASTERN ONTARIO

Lake Ontario

Lake Ontario has a smaller surface area than Lake Erie, but with an average depth of 83 metres, it holds almost four times the water for twice as long. So much water means not much freezes in winter, and a terrific amount of surface area is open to the weather for winter lake-effect snows. Prevailing southeast winds spare Ontario from most lake-effect snowstorms; instead the province receives climacteric benefits.

The lake warms the surrounding air during winter, and a warmer winter translates into a longer, more temperate growing season that is especially apparent on the lake's western shore, the Niagara Peninsula. Farmers can count on a warmer climate, because the lake has frozen over only twice in recorded history. Good for farmers and industry, but both nearly killed the lake with irresponsible farming practices and the dumping of industrial wastes. In the 1970s, governments came to their senses and banned phosphates and industrial dumping, and now the lake is back to hosting a game fish population.

However, Lake Ontario has a dark side. Those same waters that warm the land can also cause the foulest of weather. The eastern end of Lake Ontario is where travelling storm systems headed for the St. Lawrence meet either cold arctic air or the odd Alberta Clipper. Dreaded Nor'easters forming on the Atlantic coast have long fingers that often reach the southern end of Lake Ontario in early spring and explode into massive storms. These are ship-sinker events, and the bottom of the eastern end of Lake Ontario is littered with wrecks of unfortunate ships.

On the brighter side, the lake is deep with a clean shale bottom, and its currents are powerful movers of sand. Prince

Edward County, situated at the eastern end of Lake Ontario just west of the beginning of the St. Lawrence River, is the site of Sandbanks Provincial Park, which features miles of white sand beach and giant sand dunes. Weather drives the currents that move sand, but those storms occur far away from Sandbanks, and visitors can usually count on lots of sun and fun in the summertime.

Near Sandbanks and just east of Picton is Lake on the Mountain, a must-see place that's a bit of a mystery. This perfectly round lake, sitting 62 metres above the Bay of Quinte, has a constant outflow of fresh water but no visible inflow. Nobody knows for sure how water gets into the lake, but that aside, it's one gorgeous spot, with water the colour of turquoise flowing over an escarpment into Lake Ontario. While in the area, you might check out nearby Napanee, a town with a river that goes up and down like a yoyo. How is this possible? Because of Lake Ontario's seiche.

All the Great Lakes have a perpetual seiche of a few centimetres, though Lake Superior's seiche can get nasty during storms. Lake Ontario has a very pronounced seiche when winds blow from the southwest. Winds from that direction surge

water from Rochester, New York, across the lake to the Bay of Quinte, a distance of 110 kilometres. After the winds abate, the lake water sloshes back and forth like water in a bathtub.

The pendulum effect can go on for days and provides one of Ontario's rivers with a tide. Lake seiche causes water to move up and down the Napanee River in one hour and six-minute increments, a weird occurrence so far from the sea. In 1954, Hurricane Hazel's winds, which decimated Toronto, created a seiche in Lake Ontario that sloshed back and forth for days, causing extensive flooding and damage to shorelines. Lake seiche resemble mini-tsunamis that have, during the last century, claimed the lives of people enjoying a day at the beach or just out walking the dog, and sometimes stranded boats or left them hanging from their dockside moorings.

The Thousand Islands

The outflow of water from Lake Ontario into the St. Lawrence River is 24 kilometres wide and is historically known as the Lake of a Thousand Islands. Actually, there are 1864 islands, and, though some are tiny, many are large enough to support cottages and castles. Take one of the many boat tours, because this is beauty spot without rival—a central Ontario natural showpiece. Impressive but most awe-inspiring are the geological delights that Ontario's weird weather has wrought over millions of years.

Once a chain of mountains (connecting the Precambrian Shield to the Adirondack Mountains), the islands are remnant peaks scoured by glaciations and eroded by massive outflows of glacial water until only hard granite called the Frontenac Arch remained. Fortuitous for wildlife, because these stepping-stone islands form a bridge that enables migration across a natural barrier (the St. Lawrence River).

The biodiversity of this area—another Biosphere Reserve—is amazing. The fast-flowing waters of the St. Lawrence River provide a microclimate able to support a mix of southern and northern flora. The area is a transition zone, and here you'll find lush forests of hemlock, white pine, spruce and birch intermixed with more southerly pitch pine, red and white oak, butternut and various maples. Green forests, a thousand islands and water so clear you can see wrecked ships from the surface.

Tiny islands of granite and foul weather are a bad mix for sailing ships, and the Thousand Islands area is the final resting place for hundreds of vessels. The St. Lawrence River is a storm-cell conduit to the Atlantic and a nightmare for captains of old-time sailing ships. Having to navigate more than 80 kilometres of narrow channels was difficult enough in fair weather, but the constant storms caused many a ship to crash onto the rocks.

So many, in fact, that when some bright soul in Britain's Parliament suggested a canal be built to avoid the region, all the ship owners threw in with the idea. Proposed after the War of 1812 to offer passage of goods safe from hostile American forces, the canal idea gathered momentum, and in 1816 the Brits sent Joshua Jebb, a Lieutenant of Royal Engineers, to survey a route from the Ottawa River to Kingston. With some modifications, it eventually became the famous Rideau Canal.

The Rideau Canal

Under the supervision of John By, a Colonel of Royal Engineers, construction of the great canal began in 1826 at a place on the Ottawa River called Bytown—later renamed Ottawa. His plan: connect a string of lakes called the Rideaus, thereby enabling ship and barge passage all the way to Kingston, a distance of some 202 kilometres. By's workforce of diggers and stonemasons numbered several thousand, and he figured two years to completion. A doable plan, except he failed to take into consideration Ontario's weird weather and the presence of anopheles (malaria-carrying) mosquitoes. Colonel By's two-year project dragged on to become five and cost more than 1000 lives.

They called it lake or swamp fever or ague—a common ailment at the time in southern Ontario that was actually a mild form of malaria transmitted by mosquitoes. Disease and accidents plagued Colonel By's diggers and masons for the entire construction: malaria killed as many as 500; smallpox, dysentery, pneumonia and accidents killed another 500, or so.

Area weather ran from sizzling summer heat to bone-snapping winter cold, with lots of precipitation from passing tropical air masses in conflict with arctic air. It rained a lot, causing river and lake levels to go up and down. At a place called Hogback Falls, they built the largest dam (at the time) in North America four times because storm flooding caused its collapse three times. The original contractor, a Walter Fenelon, built the dam twice. By's engineers tried once and were about to have another go, when some bright lad noticed that the original cofferdam constructed by Philemon Wright, the founder of Hull, Québec, had withstood every watery assault. Colonel By wisely decided to extend it across the river and pack the sides with stone. The dam is still there and will be for many centuries to come.

Colonel By had excellent contractors, and one of the best was John Redpath, who built Montréal's Notre Dame Cathedral. The Colonel assigned Redpath the task of building the dam at Jones Falls, an architectural marvel even by today's standards, but the area was swampy and alive with skeeters carrying the dreaded swamp fever. Redpath had upwards of 300 workers, and every one, including the camp doctor, came down with malaria—many died.

By November 1831, the Canal was essentially complete. On May 24, 1832, Colonel By, his family and some fellow officers took a five-day voyage from Kingston to Bytown. The Canal was officially open for business.

Sadly, the Colonel never got his due. Some months later, he was hauled back Britain to face a parliamentary inquiry regarding the project's expenditures. While cleared of wrongdoing, Colonel By never received recognition for his tremendous feat of engineering and died four years after the Canal opened.

The Rideau Canal is now a major Ontario summer and winter attraction—a boater's dream that in winter becomes the world's longest ice-skating rink. Designated a UNESCO World Heritage Site, its year-round use by residents and visitors alike is a fitting tribute to a man who braved weird weather, malaria and a skinflint British parliament to create a marvellous inland waterway.

Significant Ontario Weather Events

*It is best to read the weather forecast
before we pray for rain.*

–Mark Twain

ONTARIO WEATHER DISASTERS

It Was a Bad Century

November 7–13, 1913—Lakes Erie and Ontario
A cyclone eased out of the Ohio River Valley, gathered strength from the warm waters of western Lake Erie and headed for the St. Lawrence River like a giant bowling ball rolling for a strike. Winds topped 140 kilometres per hour; 34 ships and 270 crewmembers were lost on Lakes Erie and Ontario. Called the Black Sunday Storm by old-time mariners, it remains Ontario's worst maritime disaster.

July 29, 1916—Northeastern Ontario
A violent thunderstorm rolled over the mining towns of Cochrane and Matheson and was initially thought heaven-sent, because surrounding forests had been bone-dry for weeks. The storm lashed the area with thunder and lightning, but the rain didn't last. When the skies cleared, multiple plumes of smoke foretold a disaster. Lightning had started fires, lots of them. They eventually crowned and destroyed both towns, killing 233 residents.

July 20, 1919—Biscotasing
It was Ontario's hottest day, at 42.2°C. This northern Ontario logging community on the shores of Biscotasi Lake had no asphalt roads to melt, no metal girders to warp and its mostly male residents suffered the day either immersed in the lake or drinking whiskey in icehouses. Bisco, as residents call it, is where the legendary Englishman, Archie Belany (better known as Grey Owl) learned to hunt, trap, lie and drink whiskey.

June 26, 1930—Brockville

The *John B. King* was the largest drilling platform in Canada. Forty-two metres in length and built entirely of wood, she looked more like a big barn than a vessel. On the afternoon of June 26, while engaged in deepening the navigation channel of the St. Lawrence River, the *John B.* was overrun by a summer storm and struck by lightning. The lightning strike detonated her already-placed charges on the riverbed, which in turn exploded her onboard store of dynamite. The blast disintegrated the wooden vessel, tossed bits and pieces over a wide area and killed 32 of the ship's complement of 43 sailors.

December 29, 1933—Algonquin Park

A weather-reporting station in Algonquin Park recorded a temperature of −45°C, Ontario's coldest day. Weather stations all across Ontario reported extreme low temperatures, and for only the second time in recorded history, Lake Ontario froze solid—an opportunity not missed by those few intrepid individuals engaged in smuggling liquor to the U.S. side, as evidenced by the foot-rutted trails that soon crisscrossed the lake's snowy surface.

July 5–17, 1936—Manitoba and Ontario

The deadliest heat wave in Ontario's history owed its existence to a strong high-pressure ridge that set up on America's west coast at the end of June 1936. That ridge had muscle and pushed dry, hot air northeastward across America's drought-stricken Midwest, the so-called dustbowl states. Poor farming practices had turned once fertile prairie lands into lifeless hardscrabble, susceptible to wind erosion and solar heating. Desert lands added more heat to the already hot air. Nothing to worry about—people down the line knew summer heat waves like old friends. A few nights sleeping on the porch was only a minor inconvenience and considered a right of summer. Except the summer of '36 would be different, and in the midst of the Great Depression, people across the U.S. and Canada suffered record heat for a record number of days—5000 lost their lives.

But the heat would not go away, and those few nights sleeping on the porch turned unbearable. People took to basements or found no sleep at all. As the heat wave stretched into the Great Lakes states, it withered crops and baked fruit ripening on trees. In Ohio, temperatures soared to 43°C, and the heat just kept sizzling as it moved toward the U.S. east coast, where the states of New York and New Jersey suffered record temperatures and many fatalities.

Along the way to the east coast, the great heat wave reached up into Manitoba and Ontario and soaked up water from the Great Lakes. For two weeks, temperatures stayed at 44°C, and the humidity made it feel much hotter. The persistent high temperatures left almost 800 Canadians dead from heat exhaustion and another 400 from heat-related factors, including drowning while seeking refuge in rivers or lakes. According to Environment Canada, the heat was so intense it twisted steel rails and bridge girders, buckled sidewalks and melted paved roads into sticky goop.

July 8–10, 1936—Toronto

The temperature in Toronto soared to 40.6°C for two days. On July 10, it peaked at 41.1°C. Residents sought refuge at beaches, in movie theatres with air conditioning and the countryside. But most had to tough it out, and 270 died from heat exhaustion. Record high temperatures in a city that two years earlier had endured a winter of record lows.

July 11, 1936—Atikokan

The temperature peaked at 42.2°C, tying the July 20, 1919, Biscotasing record. Residents sought refuge from the heat in lakes, rivers, basements, icehouses, abandoned mine shafts and places selling cold beer.

July 13, 1936—Fort Frances

Temperatures peaked at 42.2°C, again tying the record high. Forewarned by people in Atikokan, the residents of Fort Francis sought refuge from the heat in the same icehouse manner as the folks of Atikokan.

June 27, 1946—Windsor

A tornado roared across the Detroit River into Windsor, destroying or badly damaging 400 homes, killing 13 residents and injuring hundreds. Property damage to rural areas was extensive, farms and outbuildings were damaged, orchards torn up and woodlots flattened.

September 6, 1947—Gooderham
Fall Fair day in a neighbouring Haliburton town meant very few Gooderham residents were at home to witness the tornado that roared along the shore of Pine Lake. The twister uprooted trees, smashed the railroad station, destroyed a few homes, overturned boxcars, wrecked the Agricultural Centre and headed into the forest, fortunately leaving most homes and the business section undamaged. There exists a story about five people yanked from their home and deposited alive, kicking and stark naked, in the treetops. A humorous image but unconfirmed and of dubious occurrence.

May 21, 1953—Sarnia
Having already devastated the American side of Lake Huron, an F3 tornado crossed the St. Clair River from Michigan, intensified into an F4, and smashed through Sarnia, killing four and injuring 40. Estimated damage to Sarnia exceeded $15 million, with 500 left homeless. Moving northeastward toward London, it caused extensive rural damage and slowly began to dissipate. Passing just north of London, the twister mangled the small community of Nairn, moved toward Stratford and dissipated just east of that town. Michigan and Ontario both suffered terrible losses from the Sarnia Tornado; in all, it killed six, injured 60 and destroyed more than 800 homes.

October 14–15, 1954—Toronto
Hurricane Hazel caused extensive flooding from Lake Simcoe to Toronto and from the Niagara region to Windsor. Toronto, the hardest hit area, saw 7000 people left homeless and 81 killed. Hazel, an almost-spent, north-tracking cyclone, collided with a southeast-tracking low and became extratropical (which means it didn't form over water). Hazel dumped more than 210 millimetres of precipitation on already rain-soaked areas of Toronto, causing local rivers to flood their banks in

SIGNIFICANT ONTARIO WEATHER EVENTS

a spectacular fashion. Damage was catastrophic, and the estimated cost of the storm was more than $1 billion.

July 25, 1965—Erieau
It rained fish in this small community beside Lake Erie. A waterspout wandered in off the lake and dropped a few tons of water onto the main street, along with a mess of fish. After mangling a few cottages, the water twister returned from whence it came, leaving behind a street filled with flopping bluegills and perch. There are many historical reports of storms raining down fish, toads and frogs. There are even reports of larger creatures—horses, cattle, pigs and even people—being hurled into new locales via twisters.

July 13, 1973—Brighton
The storm began early in the afternoon over Lake Simcoe and moved southeastward into Lake Ontario. Here it developed into two separate systems, one of which spawned a tornado.

Brighton was the iconic Ontario town: squeaky clean, maple-lined streets and smart-looking houses with long verandas and expansive backyards. All that ended at 7:30 PM, when a borderline F1 tornado ripped up the main street like a giant buzz saw. In less than a minute, the maple trees lay in broken heaps, house roofs were gone, vehicles lay tossed about and the town's beautiful church had lost its steeple. They rebuilt Brighton, but maple trees no longer shade the streets, the verandas weren't replaced and those expansive backyards are now flagstone patios.

May 31, 1985—Southwestern Ontario

A line of thunderstorms crossed into southwestern Ontario from Michigan and spawned 11 tornadoes. The storm system extended in a horseshoe shape across southwestern Ontario and accounted for 12 fatalities and more than $150 million in damages. The first tornado touched down at Leamington and the last at Alma, where the system crossed back into the U.S. and continued to rampage. The major damage from this system was to the city of Barrie on Lake Simcoe.

July 18, 1991—Pakwash Forest

Approximately 1500 square kilometres of northwestern Ontario forest was flattened by non-tornadic winds and downbursts. The damage path, roughly 20 kilometres wide and 75 kilometres long, was later thought by some area residents (who had sought refuge in those ubiquitous icehouses) to be caused by either a UFO or Babe, Paul Bunyan's blue ox. Estimated number of trees destroyed: 30 million.

July 14, 1993—Windsor

Humidex values at Windsor soared above 50°C, the highest reported in Canada to date. Residents flocked to beaches, movie theatres and places selling the coldest beer. Called the "nation's armpit" by residents, Windsor suffers more high-value

humidex days than any city in Canada. If you're considering a career in the air-conditioning business, this is your spot.

April 20, 1996—Grey and Wellington Counties

Two F3 tornadoes touched down—the first one in Grey County near Williamsford, where it tracked 40 kilometres southeast of Owen Sound. The second formed in Wellington County southeast of London and tracked 60 kilometres to a point just southwest of Barrie. Both did extensive damage to property, but luckily, no lives were lost.

Both tornadoes were interesting for their early-in-the-season arrival and the topography over which they travelled. The regions are hilly with high elevations, and neither had experienced a tornado before. The Grey County tornado was the most interesting, because it pushed a line of weird weather ahead of it like a bulldozer. Weather in front of that tornado featured dark skies with rain and hail, while behind it, the sun shone in a cloudless sky.

January 8, 1998—St. Lawrence River Valley

A warm, elongated low-pressure air mass, moving up from the south and into the St. Lawrence River Valley, wedged into a belt of cold arctic air and shoved it eastward, while releasing huge amounts of precipitation. Supercooled by the frigid air underneath, the below-freezing rain stuck onto any cold surface, a weird weather process that continued for five days. Canada's worst ice-storm disaster affected millions of people, damaged or destroyed thousands of farms, smashed flat whole forests and toppled a tremendous number of transmission pylons, cutting power to eastern Ontario, most of Québec and some northeastern U.S. states. In all, this ice storm killed 34, injured 945 and caused almost $7 billion in damage.

June 2, 1998—Norwich

A cold front swept into southern Ontario from the west, forming severe thunderstorm activity as it moved eastward. Supercells in this front caused an outbreak of mostly small tornadoes, but two—an F2 and an F1—caused extensive damage to rural areas and various communities. The F2 smashed the small town of Norwich, and the F1 mangled the town of Dunnville situated on the shore of Lake Erie. Fortunately, no fatalities occurred.

January 13, 1999—Southern Ontario

The Great Blizzard of '99 buried six Midwestern U.S. states before dumping in excess of 30 centimetres of snow onto

Ontario towns and cities. A "storm of the century," it paralyzed the entire Windsor to Québec City weather corridor and so overwhelmed Toronto that Mayor Mel Lastman called in the Canadian army to help dig out the city. Combined U.S. and Ontario fatalities numbered 73, and the estimated cost was in excess of $400 million.

September 3, 1999—Windsor

The worst weather-related highway disaster in Canadian history occurred early on the Friday morning prior to the Labour Day holiday. Ontario Provincial Police units, patrolling Highway 401 near Windsor before sunrise, reported ground fog in low-lying areas. Nothing to worry about, the sun would soon come up and burn off the fog. Except that, instead of dissipating at sunrise, the fog thickened and reduced visibility to a metre. One driver, coming over a rise and seeing the road ahead vanish into pea soup, stood hard on his vehicle's brakes. The driver following had no time to stop and smashed into the braking vehicle, effectively blocking the highway. One by one, vehicles sped into the fog only to be added to a massive pile-up that killed eight, seriously injured 33 and destroyed 84 vehicles.

July 1, 2001—Kapuskasing, Ontario

Three centimetres of snow fell on the town's Canada Day celebration, prompting a massive snowball fight and cancellation of a volleyball tournament and popular rib feast.

March 17, 2003—Barrie

Although not Ontario's worst weather-related traffic pile-up (no deaths resulted), it was certainly the longest, involving 20 kilometres of Highway 400 near Barrie. Thick fog and slippery road conditions contributed to a pile-up of more than 200 vehicles and shut down a major highway for days.

SIGNIFICANT ONTARIO WEATHER EVENTS

July 15, 2004—Peterborough

An intense thunderstorm deluged the Peterborough area in the early-morning hours. Official rainfall totals ranged from 100 millimetres at the airport to 240 millimetres at Trent University, with most of the accumulation falling in less than five hours onto the mayor's house.

November 9, 2005—Hamilton

Ontario's weirdest weather day began around noon, when a line of thunderstorms formed over Lake Huron, moved over the province and caused extreme weather variations. Windsor enjoyed temperatures of 20°C, while Ottawa endured freezing rain. Barrie got snow; Hamilton saw a rare off-season F1 tornado that tracked for seven kilometres, flipped vehicles, ripped roofs and smashed a school gymnasium, injuring a few students. The Hamilton tornado was only the third known late-season tornado to touch down in Canada. Oddly, those other late-season tornadoes also hit Ontario: Leamington on November 29, 1919, and Exeter on December 12, 1946.

Summer of 2006

This was the worst and weirdest summer-season weather since record keeping began in Ontario in the early 1900s: 23 recorded tornadoes and three massive windstorms. That summer of weirdness cost Ontario taxpayers and insurance companies more than $100 million and at various times cut power to huge sections of the province.

SURVIVING NATURAL DISASTERS

Assume the Worst and Plan Accordingly

Ontario insurance companies pay out millions of dollars every year in compensation for losses caused by natural disasters; costs to provincial and municipal governments for the restoration of damaged infrastructure can run into many more millions. Monetary costs aside, the human toll from a disaster can be devastating to families and Ontario communities. Don't let a natural disaster catch you unprepared.

Here are some tips for surviving a natural disaster:

- Strengthen buildings and roofs. Simply tying down a roof with cables will dramatically increase your chances of survival.

- Install hail- and fire-proofed shingles, and install shutters to prevent glass blowout during storms.

- Purchase a gas- or diesel-powered electric generator.

- Install backflow valves on the house sewer system to prevent overflows in case of flooding.

- Secure furniture and appliances to walls and outdoor furniture to solid ground.

- Build a safe room if you're in an area of high tornado incidence.

- Keep a clear area around your house, and trim overhanging branches.

- Keep drainage and streambeds clear of refuse.

- Have adequate and proper insurance, keep it current and visually document your house and its contents so that if you have to make a claim, you have proof of your possessions. Stash that documentation in a safety deposit box; don't leave it in your house where it might be destroyed along with the rest of your household goods.

Tornadoes

Seeking tornadoes is looking for devils; go unprepared, and the devils will search for you.

–A.H. Jackson

DOING THE TWIST(ER)

Measuring the Damage

Nothing frightens like a twister. Hurricanes are larger and cause more damage, but they're not in-your-face dangerous like tornadoes. Meteorologists give names to hurricanes; twisters are numbered 0 to 5 on an F scale—the Fujita scale.

This scale—developed in 1971 by T. Theodore Fujita of the University of Chicago—measures the damage to human-made structures to rate the intensity of a tornado.

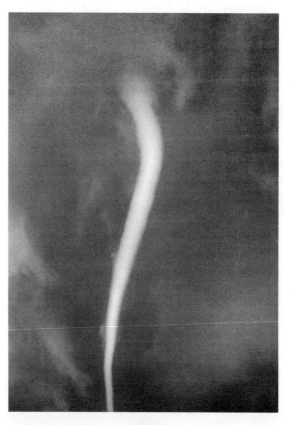

TORNADOES

Wind speeds are based on estimates after the damage has been examined. Sixty-seven percent of tornado-related deaths occur during F4–F5 storms and 29 percent during F2–F3 tornadoes.

F0: light winds, 64 to 116 kilometres per hour—light damage to structures; tree limbs broken; 28 percent of all tornadoes.

F1: moderate winds, 117 to 180 kilometres per hour—moderate damage; mobile homes pushed off foundations; cars blown off roads; 39 percent of all tornadoes.

F2: considerable winds, 181 to 252 kilometres per hour—considerable damage to structures; roofs torn off frame houses; trees uprooted; 24 percent of all tornadoes.

F3: severe winds, 253 to 330 kilometres per hour—severe damage; houses levelled, vehicles tossed around; six percent of all tornadoes.

F4: devastating winds, 331 to 417 kilometres per hour—devastating damage; structures disintegrated or blown away; two percent of all tornadoes.

F5: incredible winds, 418 to 509 kilometres per hour—absolute damage; houses and cars flung through the air; incredible phenomena; less than one percent of all tornadoes.

In Canada, most high-number F-scale twisters occur in southern Ontario, Alberta, Saskatchewan, Manitoba and southern Québec, places with extreme weather variables—our tornado alleys. Ontario is Canada's "tornado central" and sees about one-quarter of all Canadian tornadoes, mostly because of its proximity to the Great Lakes and the Ohio River Valley.

During the spring and summer months, warm soggy air from the Gulf of Mexico speeding north through the Ohio River Valley often collides with northeast winds from central Mexico. If this odious pairing should encounter cold southeast winds blowing from the Rockies or the Arctic, the result can be a dramatic wind shear.

Wind shear is a meteorological term used to describe winds that blow in different directions at different altitudes. When that northerly moving air mass from the Gulf of Mexico encounters arctic cold, it rises and is pushed by wind shear into a horizontal rotation. If strong wind shear pushes one way at the bottom and another pushes the opposite way at the top, it will create a horizontally spinning snake-like wind tube.

Ground heat, or heat rising from Great Lakes water, will gravitate toward the low pressure created by the spinning tube and pull it down onto the earth, where it sucks warm, moist air high into the atmosphere. On the way up, that moist air condenses into water droplets and a thunderhead is created.

However, that air can rise even farther, sometimes to 18 kilometres, where those water droplets freeze into ice crystals. These high, powerful storm clouds—called supercells—can bombard the earth with hail and lightening and can spawn single or multiple funnel clouds.

Tornadoes are fast-moving horrors, like atom bombs in motion. Fortunately, southern Ontario does not see as many as the U.S., where tracking them has become something of a sport, but Ontario does get them on a regular basis. Ontario sees most twister action in a row of counties from Lake Erie to Lakes Huron and Simcoe, a major tornado alley.

Be leery of the supercell. It hides in a storm cloud that has you staring awestruck and thinking: "Gee, that looks like trouble." If you think that, it probably is, and the trouble can be horrendous: hail, high winds, lightning, downbursts and tornadoes from single or multiple storm cells. Canada's Arctic and southern Ontario are an intrinsic part of the tornado-forming equation; the former supplying the cold, dry air, the latter kicking in the warm, moist air rising off the Great Lakes—especially from a lake called Erie.

Environment Canada records about 80 Canadian tornadoes annually, with about 15 to 20 touching down in Ontario—most in the Great Lakes basin. But those figures are vastly understated, because many twisters go unobserved or unreported. A tornado's life on ground can be short, and those worthy of reporting are few in number. A twister touches down, tears up a farmer's woodlot or pasture and maybe tips a cow or a shed—not a big deal, even if the farmer sees it happen. Twisters are usually only reported by meteorologists, police and hospitals, and then not always. A more realistic count of Ontario tornadoes would be around 40, with most occurring in the Great Lakes basin during the months of June and July.

Here is the required weather formula for the creation of a tornado. Take a mass of warm, dry, continental tropical air from Mexico or the U.S. southwest and toss in some cold, dry continental polar air from Canada's subarctic regions. Mix well with a splash of hot, wet maritime tropical air from the Gulf of Mexico. Always a volatile mixture, this brew can become explosive around Lake Erie during the months from April to October.

The triangle formed by Lakes Erie, Huron and Simcoe plays host to travelling storm cells during the warmer months. There are times when storm traffic is so heavy that they have to line up to get into the region, and this can be troublesome.

Twister Attacks

On occasions, storm cells can form along a broad front like soldiers and spawn numerous tornados, the record being 148 with 16 on the ground simultaneously. This was the infamous American "Super Outbreak" of April 3 and 4, 1974, when tornados generated by a 30-cell family of storms chewed up portions of 13 U.S. states and parts of southeastern Ontario. A single F3 twister from this system crossed the Detroit River into Windsor, Ontario, where it smashed houses, downed trees and obliterated a curling club, killing nine members.

Eerily enough, a few decades earlier—June 17, 1946—a single F4 tornado followed almost the same path across the river, but it stayed longer, did more damage and killed 17 people. Environment Canada has posted pictures of this tornado online and they are well worth a look. One look is valued at a thousand words and might save your hide.

August 7, 1844—Galt
A minor twister knocked over trees, smashed buildings and caused one fatality, a woman—Canada's first recorded tornado fatality.

August 7, 1979—Woodstock
Tornado outbreaks are freakish nightmares, but enduring three F4 monsters simultaneously can be a mind-altering experience. On August 7, residents of Woodstock were expecting a usual late afternoon breeze. It never arrived.

Heat waves continued to waft from the streets, and the chirring of cicadas grew to an almost deafening crescendo. Around 6:30 PM, the cicadas suddenly stopped chirring, and folks sitting on porches and out walking their dogs heard a freight train coming down the tracks. A strange sound, because the trains no longer passed through Woodstock, and the tracks had been gone for years. Sensing something bad coming down the road, the dogs bolted, leaving their masters to ponder what sort of monster was at their gates.

'Twas a three-headed monster—three huge tornadoes touched down northeast of Woodstock, covering the entire horizon. The stuff of nightmares, and two of the biggest, both F4s, sped straight through town, chewing up everything and leaving hundreds homeless. Not only Woodstock but the entire area took a thrashing, as the twisters cut a path of destruction more than 129 kilometres long. Those tornadoes, packing winds up to 400 kilometres per hour, destroyed or damaged more than 600 homes and buildings from Stratford through to Woodstock and Jarvis. Two communities, Oxford Centre and Vanessa, were totally flattened by the Woodstock tornadoes, and damage to the area ran to more than $100 million.

Amazingly, only one person died during the Woodstock event—the driver of a vehicle blown off the road. The injured numbered around 142, and that brought in ambulances from surrounding community hospitals. Neighbours opened doors to the homeless, and a local radio station began an around-the-clock campaign to raise money, which the Ontario government matched dollar for dollar. Contributions poured in from everywhere: money, clothing and food. The warm and fuzzy feelings got even better when hundreds of Mennonites arrived to help with the clean up and stayed for the rebuilding.

May 31, 1985—Barrie

Another major tornado outbreak occurred on May 31, 1985, when 41 twisters touched down in areas around the Great Lakes, 13 of them in southern Ontario. Double bubble, toil and trouble—this witch's brew began the day before, when a strong low-pressure system swept through the U.S. plains states and pushed a warm front into the Ohio River Valley, creating a line of thunderstorm activity. The next morning found the storm gone and the sun shining, but within a few hours, a cold front emerged from that low-pressure system, and all hell broke loose…in Ontario.

The first of the monster's 41 tornadoes, a small one, touched down near Leamington (Essex county) around 8:00 AM. The next, an F2 bigger boy, touched ground near Wiarton in Bruce County, home of Wiarton Willy, the spring-forecasting groundhog. Rumour has it that Wiarton Willy did not survive that event and was replaced by a look-alike.

A little further south, near Hopeville (Grey County), an F3 churned up the countryside at 10:00 AM for a distance of 17 kilometres, and 15 minutes later, another F3 rampaged a 50-kilometre section of the Mansfield ski area.

At 10:30 AM, a small twister stirred up the town of Alliston (Simcoe County), and meteorologists tracking the storms must have thought that to be the end of them and breathed a sigh of relief. But the end for many was just the beginning for some, as two giant F4 tornadoes touched down near Barrie (Simcoe County), a city beside Lake Simcoe.

One of that pair rampaged around Arthur and Mt. Albert (Wellington, Dufferin, York and Simcoe Counties) tearing up trees, farm buildings and houses for 107 kilometres and causing four fatalities. The other headed straight for Barrie like it had a map.

The area around Barrie and Lake Simcoe is no stranger to tornadoes; every year, media report a few small ones touching down, tearing up hay, tipping cows and dissipating. Very few residents of the city had seen one of these, let alone a big twister.

All that changed a few minutes before 5:00 PM on May 31, when a monster F4 appeared southwest of town. The beast was 600 metres wide and black as sin; they know the width, because the twister ripped through a pine forest on the edge of town cutting a swath that wide. Clear of the forest, that monster twister burst into town and took out an entire block of older frame houses, causing three fatalities. Next came an industrial park, where it destroyed or badly damaged 16 factories and caused one fatality. Luckily, area power had failed some hours earlier and most of the workers had left the buildings.

Then the twister hit the Barrie racetrack; it wrecked barns, tossed around trucks and trailers and caused horses to stampede for their lives. The beast then crossed a major highway, heaving vehicles into nearby fields.

A subdivision came next, but the swath of destruction had shrunk to 300 metres. A few minutes later, at another industrial park, it smashed 11 factories before shrinking to 100 metres. The monster was progressively weakening, but it had enough power to pulverize another subdivision, causing another four fatalities. Finally, after smashing up a local marina, the almost-spent tornado turned and headed out into Lake Simcoe, where it dissipated. The marina later reported that the tornado took along 35 boats with their concrete mooring anchors. In all, eight people died, 155 were injured and damage amounted to many millions.

The storm had not finished with Ontario, and at 12:30 AM, both Wagner Lake and Reaboro (Durham & Peterborough Counties) were rampaged by a pair of F1s that did minimal damage. At roughly the same time, a giant F3 struck the Rice Lake (Northumberland County) area, while two others, an F3 and an F1, smashed into Wellington County, near Minto and Alma. A night of dread and horror for southern Ontario ended at Alma, when the massive tornado-spawning storm moved south of the border. By the end of the rampage, the international ticket price for damage totalled almost a billion dollars.

Downbursts

What goes up must come down—and when a developing supercell begins to vacuum warm air into its column, it goes up fast and becomes extremely cold. At the top, this heavy mass of icy air shoots out from the column and drops toward the ground like a lead weight.

Some old-timers call this phenomenon a "plow wind," because that's what it does—it plows into the ground and blasts across the surface while making a noise like a freight train…a tornado noise. These can be narrow (micro) and blow your house to smithereens, or wide and flat (macro), in which case they only tear off roofs and topple trees. Both are extremely hazardous to boats, ships, aircraft, hydro transmission towers and forests.

On May 12, 2000, a downburst that included hail, a rain deluge and violent 140-kilometre-per-hour winds struck Niagara-on-the-Lake, smashing 200-year-old trees and houses and severely damaging an 1832 historic inn.

Derechos

Spanish for straight, as in straight winds, a derecho is a violent windstorm associated with a band of fast-moving thunderstorms that can feature one or multiple downbursts. The difference between derechos and normal windstorms is sustained wind speed (in excess of 92 kilometres per hour)—your basic, non-gusting, comma shaped, non-tornadic event. Still, these are horrors on a massive scale, a line of atmospheric bulldozers that will smash everything in a kilometres-wide path. One or more downbursts can help increase derecho straight-line winds to incredible speeds of up to 250 kilometres per hour in extreme cases. Tornadoes can spawn from these extreme derechos, but unless they spin out, they're just more fuel for the fire, considering the ongoing damage.

In Ontario, derechos tend to form in the northwest during periods of extreme heat and along stationary fronts. Derechos are divided into three types. A serial derecho is a massive squall line of low-pressure supercells about 250 kilometres long; a progressive derecho is a smaller squall line able to travel long distances; and a hybrid derecho is (as you might guess) a compact combination of serial low-pressure cells and far-travelling progressive cells.

During the 1993 "storm of the century" that mangled the entire east coast of North America, a serial derecho forming ahead of an exploding low-pressure storm cell lashed Florida and Cuba with straight-line winds exceeding 160 kilometres per hour. That derecho spun off tornadoes like an automatic weapon, causing horrendous damage and numerous fatalities.

May 31, 1998—Lakes Huron and Erie

A squall line of thunderstorms moving eastward over Minnesota during the early morning hours enclosed a massive stationary supercell storm. With its character changed into a comma-shaped derecho, the storm headed across the Great Lakes with sustained winds in excess of 140 kilometres per hour, spawning tornadoes and spewing lightning bolts like bullets from a machine gun. This small, extremely low-pressure, hybrid derecho knocked down everything in its path before smacking into Lakes Huron and Erie, like a boot into a puddle, causing lake surge and considerable damage to shorelines.

July 4, 1999—Ottawa

A band of thunderstorms formed near Fargo, North Dakota. They featured winds gusting to 180 kilometres per hour that caused extensive damage. As the storm moved east toward Minnesota, its character changed, and its winds turned straight-line with sustained speed, a derecho. In Minnesota, the derecho overran the Boundary Waters Canoe Area and Superior National Forest, both packed with people enjoying the American Independence Day holiday. Sustained winds of 140 kilometres per hour killed one camper, injured 60 and knocked down an amazing 25 million trees.

During the early afternoon, the storm passed into northwestern Ontario. There, its sustained wind speed increased to 160 kilometres per hour as it headed east toward Québec at about 60 kilometres per hour.

Approaching Ottawa, this progressive derecho spawned numerous tornadoes and featured massive lightning strikes estimated at 6000 per hour. Narrowly missing the capital, it moved over Montréal during the early hours of July 5, headed into the Québec Townships and continued toward Maine, where it dissipated. In all, that derecho killed four, injured 70, caused 700,000 homes and businesses to lose power and cost more than $100 million.

October 25, 2001—Southwestern Ontario

Derecho winds of 100 kilometres per hour lowered water levels in the Detroit and St. Clair Rivers by more than 1.5 metres and caused havoc with shipping. Lake vessels on the water had to drop anchor, and those about to sail remained at dock for days. Ships waiting to traverse the Welland Canal were also blocked, and the overall cost of that very windy day ran into the millions.

July 17, 2006—Southern Ontario

On July 17, the upper U.S. Midwestern states and most of Ontario were in the grip of a heat wave that spawned a derecho of grand proportions. It began when a low-pressure squall line formed across the Michigan peninsula and moved over Sault Ste. Marie into northwestern Ontario. The squalls seemed to like the area and hung around, while the storm matured into something bigger and meaner, with the characteristic comma shape that is the radar tell of a derecho storm. A progressive derecho, as it turned out, and with sustained winds in excess of 100 kilometres per hour, it ran straight east all the way to Québec's Laurentian mountains, smashing towns, farms and a few million trees along the way.

SAFETY FIRST

Don't Be Arrogant

So there you are, driving along, minding your own business and thinking you need new windshield wipers, because the ones you have barely kept up with that rainstorm. Had it lasted longer you might have been forced to pull over, but it passed, and you're all right. All that remains is a weird-looking rotating cloud dragging behind the retreating thunderstorm. An orphan, and when it moves in your direction, you're more amused than concerned.

Big mistake. You should take immediate action and find a right-angle road away from that cloud. Do it quickly, because that orphan is a "wall cloud," the place where tornadoes spawn. Dangerous even without a tornado, it can bring along a phenomena storm chasers call "the bear cage," a micro downdraft that can flatten your vehicle like a pancake.

All thunderstorms are intrinsically dangerous, and supercell storms are off-the-scale deadly. When driving in an area of thunderstorm activity, tune your vehicle's radio to a local station: a "tornado watch" alert means a possibility; a "tornado warning" means one is on the ground, having been observed or detected by Doppler radar. Warnings should have you seeking shelter immediately, and that does not mean simply pulling off to the side of the road. If a tornado overruns your vehicle, you will not survive, end of story. Step on the gas and get far away, while keeping an eye peeled for twisters. If you spot one, take immediate evasive action. Don't become mesmerized and miss a turn away from the beast. Very important: tornadoes do not follow a straight line. Employ extreme caution when taking evasive action and keep alert for multiple twisters.

TORNADOES

If for some reason evasive action is not possible, get out of your vehicle and find substantial shelter. If no shelter is available, find a low place like a ditch and protect your head. Don't seek shelter in highway underpasses—their configuration will increase wind speed and send you on a one-way trip to Oz.

If weird weather threatens and you're at home, keep the radio tuned to a local station and have a safe spot already prepared. Basements are best, but if that's not possible, choose a main floor bathroom or closet without windows. Stay away from windows, and don't open them. The air pressure will not blow up your house, but it will explode glass.

Plan and Practice

Make sure everyone in the house knows what to do and where to go, quickly. Conduct a family drill during both day and night, the latter being important because that's when most tornado fatalities occur. Get the drill down pat and remember to include pets. Keep an emergency kit handy—flashlight, candy bars, water—and don't forget your cell phone.

If you reside in a trailer home and spot a tornado on the horizon, fall onto your knees and pray. Just kidding, but you must have a safe spot away from that trailer. If you stay and are overrun, you will not survive the experience.

Contrary to popular belief, tornadoes have no affinity for trailer homes and trailer parks; it's all about design and construction. Built light to facilitate towing, house trailers take on the characteristics of a kite when overrun by tornadic winds. Unless you want a trip to Oz, get out quick and not to your neighbour's trailer. Do not panic, just run to the spot you have predetermined will withstand a twister. If you're in a resort park with community bathing facilities, go there and hunker down in a corner away from any windows.

Lightning

Thanks to modern science, we now know that lightning is nothing more than huge chunks of electricity that can come out of the sky, anytime, anywhere, and kill you.

–Dave Barry, syndicated humour columnist

LIGHTNING

A CHARGED AFFAIR

Don't Fry and Die

Lightning kills an average of 16 Canadians every year, with most fatalities occurring in Ontario. But that's not the whole story, because lightning wounds and maims hundreds and most strikes go unreported. Lightning can cripple, blind and mentally incapacitate its victims and can find you almost anywhere at anytime. Lightning can enter your home through a telephone line and fry your brain or through plumbing and boil you alive while you're having a bath.

So: do not take a bath or shower during thunderstorms, do not talk on a landline telephone and do not stand near a window to watch the pyrotechnics. If you live in a newer house with PCV pipes, get a lightning arrester installed, because that type of piping will not ground a strike.

Especially during the summer months, lightning can be a supreme hazard to boaters, because the vessel is usually the highest object on the water. Fibreglass and wood sailboats are especially vulnerable—their masts are high and their construction materials are unsuitable for grounding electrical charges unless the hull is wet from rainfall. Lightning will cause catastrophic damage to wood or fibreglass boats and their occupants, so stay clear of storms on the water.

An automobile or truck provides excellent protection during a thunderstorm, but not because of the rubber tires—the reason most people think. In fact, it's the "skin effect," which occurs when lightning strikes a metal car or truck body and travels over its surface—the roof and sides save you from terminal zap.

The same is not true for riding lawnmowers, golf carts, bikes or motorcycles. If a thunderstorm catches you riding one of

LIGHTNING

those, stop and bail out quickly. Drivers in vehicles with fibreglass bodies should pull to the side of the road and advise all occupants to sit with hands folded until the storm passes. This is a good idea in any vehicle: keep your hands folded, do not touch any metal objects and keep the windows closed. Lightning striking an automobile will usually pick the radio antenna or a windshield wiper to exit. Either will temporally blind the driver with a bright flash, so it's best to pull over until the storm passes.

LIGHTNING

Chances are good you'll never be struck by lightning. According to Environment Canada, you have a 1 in 4000 chance of being zapped in your lifetime. Not bad odds, but some Canadians, the author included, have defied those odds and been struck numerous times, so why gamble; it's always better to be safe than crispy.

Scientists are not exactly sure why thunderclouds create lightning, but most votes are for a forceful splitting of electrons by fast-forming ice crystals. Whatever the reason, thunderclouds like to create lots of lightning in southeastern Ontario. During early spring and summer, thunderstorms in this area can strike ground approximately four times every second.

On the lightning hot-spot scale, the safest city in Ontario is Thunder Bay, with a mere 36 strikes per square kilometre. Farther south, Windsor and surrounding areas receive a fantastic 251 strikes per square kilometre. That's right. That's also Ontario's major tornado alley, and during summer storms it can seem like the whole area is either whirling in wind or dodging lightning bolts. It can be a dangerous part of Ontario if you fail to take precautions, one of which is staying tuned to a radio station for weather alerts. Locals tune to the Detroit station, claiming that Environment Canada's $400 million system only seems to issue warnings after the threat has passed.

A Bolt from the Blue

Except for dark clouds on the horizon and a distant rumble of thunder, the storm has passed. Golfers take to their carts, carpenters return to hammering, gardeners dig and dog walkers hit the sidewalks—an oft-repeated scenario that sometimes leads to tragedy. Meteorologists call it "clear-air lightning" because it seemingly strikes without source.

After a storm, blue sky and sunshine return, and suddenly, without any crack of thunder, a golf cart or tree explodes. It's a rare phenomena and little understood, but every year it claims a few more victims. Contrary to the 95 percent of lightning strikes that are negatively charged, bolts from the blue are positive energy originating in the anvils of high-altitude altocumulus clouds. Because of this high origin, targets 20 or more kilometres away can be struck at any time and in clear weather. If that passed-over storm cloud on the horizon has a top shaped like a blacksmith's anvil, you should put off going outside for a few more minutes.

Dry-air Lightning

This phenomenon is related to clear-air lightning, but is less rare and kills far more people. The storm has passed, the sky is still overcast, but the rain has stopped and a pleasant dry wind has people resuming their normal activities. Then, seemingly from out of nowhere, ZAP!

But how can that be? No rain and air dry as toast. Doesn't air have to be wet to conduct electricity? Normally yes, but on occasion, tall objects will remain electrically charged after a storm has passed and can sometimes be observed sparking out a blue flame called St. Elmo's Fire, named for the patron saint of sailors, St. Erasmus of Formiae.

A word of warning: if you should see this phenomenon, hit the dirt and make like a pancake. Unable to release its negative charge to the atmosphere, that tall sparking object is searching out a suitable ground, which could be you walking the dog. Tall trees, church steeples or telephone poles—it can come from anywhere, so be aware.

Super Bolt!

Lightning is basically a kid shuffling feet on granny's thick carpet while chasing his siblings around. Same spark, just much bigger, the average being around 30,000 amps.

However, on occasion, lightning amperage can rise dramatically to an astounding 300,000-plus amps. These super bolts are six or seven times hotter than the sun and very destructive. If struck by one, you are toast.

Super bolts have struck baseball fields and killed players; they have fried golfers, set barns and houses on fire, burned forests and knocked down countless power transmission towers. They seem to like transmission towers, and smashing them costs Canadians $10 million every year. Thankfully, super bolts are rare, and your chances of being struck by a normal bolt are far more likely, as are your chances of surviving.

Super bolts carry a positive rather than a negative charge and come from those flattened tops of altocumulus clouds that meteorologists call the anvil. This is a place of great atmospheric activity caused by warm, moist air rising in the great cloud's updraft and colliding with cold air at the top. How cold? To put this into perspective, a satellite with temperature-measuring abilities passed over tropical cyclone Hilda east of Australia in 1990 and measured temperatures at a top altitude of 18.9 kilometres. The result was a fair dinkum −102.2°C, a temperature difference of 97.2°C between top and bottom.

At such low temperatures, water droplets freeze as hard as titanium and release energy that's off the scale. Positive lightning, or super bolts, can strike the Earth far from any storm activity, and for that reason they're called "out-of-the-blue strikes." The storm is over, lots of sunshine, life is good again—and then ZAP, all they find of you is a smoking ear.

Ball Lightning

During World War II, ball lightning was the bane of submarine engineers. At the call of, "Dive, dive, dive!" the engineer had to rapidly switch from diesel motors to batteries. The switches required simultaneous flipping to prevent arcing, and if not accomplished correctly, a luminescent ball of fire would pop out from under the batteries and burn the legs of careless engineers. In the same war, but higher up, fighter pilots constantly reported seeing luminescent balls they called "foo fighters."

Scientists of the day proclaimed the luminescent balls in submarines to be simply figments of the imagination caused by stress. As for the foo fighters, well, those were bits of super-heated engine grease burning at high altitudes. Fact is, scientists had no idea about those luminescent balls of fire, nor did they believe in their existence. Things change over the years, and today's scientists are now believers—but still have no idea what causes the luminescent balls of lightning. Theories abound, but all are guesses because finding an "*au naturel*" ball of lightning to actually study is all but impossible.

Many years ago, I experienced ball lightning while in command of a large propeller-driven aircraft. About the size of an orange, that snapping ball of electricity terrorized me and my co-pilot

for more than five minutes before we determined it was drawing power from the propellers. Off went the engines and out through a thick glass window went the ball. Moral of my story: don't look out windows at any atmospheric electrical phenomena, and be especially leery of bright, luminescent balls.

LIGHTNING

BE LIGHTNING WISE

In 1752, Benjamin Franklin, U.S. political activist and science dabbler, flew a kite into a thunderstorm during an experiment to explain static electricity. One year later, after reading of Ben's experiment, a Russian scientist, Professor Georg Richmann, recreated the kite 'n' key adventure and launched it from the doorway of his house into a violent storm. But, unlike Ben, who had succeeded in getting only sparks, Professor Richmann got a ball of lightning that blew out his doorway, leaving him dead as the proverbial doornail. Moral of that story: never cast your line into a thunderstorm, or you might be "the catch of the day."

Avoid this fate, and be lightning wise. To help you out, here are a few tips from Environment Canada. Start by learning the 30-30 Rule. Take appropriate shelter when you can count 30 seconds or less between the lightning flash and thunder. Remain sheltered for 30 minutes after the last thunderclap.

If You're Outside

- Stay a safe distance from tall objects, such as trees, hilltops and telephone poles.

- Avoid projecting yourself above the surrounding landscape. Seek shelter in low-lying areas, such as valleys, ditches and depressions—but be aware of flooding.

- Stay away from water. Do not go boating or swimming if a storm threatens, and get to land as quickly as possible if on the water. Lightning can strike the water and travel some distance from its point of contact. Do not stand in puddles, even if you're wearing rubber boots.

- Stay away from objects that conduct electricity, such as tractors, golf carts, golf clubs, metal fences, motorcycles, lawnmowers and bicycles.

- Avoid being the highest point in an open area. Swinging a golf club or holding an umbrella or fishing rod can make you the tallest object and a target for lightning. Take off shoes with metal cleats. If you're caught on a level field far from shelter and feel your hair stand on end, lightning may be about to hit you. Kneel on the ground immediately with your feet together, place your hands on your knees and bend forward. Do not lie flat. The object here is to make as small a target as possible.

- You're safe inside a vehicle during lightning, but don't park near or under trees or other tall objects that can topple during a storm. Be aware of downed power lines that might be touching your car. You will be safe inside the car, but you might receive a shock if you step outside.

- In a forest, seek shelter in a low-lying area under a thick growth of small trees or bushes.

- Keep alert for flash floods, sometimes caused by heavy rainfall, if seeking shelter in a ditch or low-lying area.

- If you're in a group in the open, spread out and make sure everyone stays several metres apart.

If You're Inside

- Before the storm hits, disconnect electrical appliances, including radios, television sets and computers. Do not touch them during the storm.

- Don't go outside unless absolutely necessary.

- Stay away from doors, windows, fireplaces and anything that will conduct electricity, such as radiators, stoves, sinks and metal pipes. Keep as many walls as possible between you and the outside.

- Do not handle electrical equipment or landline telephones. Use battery-operated appliances only.

Don't Become a Crispy Critter

Anyone struck by lightning receives an electrical shock but does not hold a charge and can be safely handled. Victims might be suffering from burns or shock and should receive medical attention immediately. If breathing has stopped, administer mouth-to-mouth resuscitation quickly. If breathing and pulse are absent, cardiopulmonary resuscitation (CPR) is required. Send for help immediately. If on a golf course, look for the course marshal or ranger; he will have a defibrillator and know how to use it.

Most folks struck by lightning survive and go about their business as if nothing happened, which explains why official records should not be trusted. Many folks who get zapped never report it. Report to whom? Strike victims usually only

LIGHTNING

tell their friends and loved ones, perhaps because official reports require proof. The only reporting comes from the police and hospitals, where the proof is usually still smoking and smelling indisputable.

Lightning is a supreme danger that dumb consensus has happening to other people. Firefighters and paramedics call those other people "crispy critters," and they are not a pretty sight. Electrocution is a painful end; avoid it by finding suitable shelter at the first indication of a thunderstorm.

One summer, while gathering info for a mystery novel, I worked as a golf course marshal, or course ranger. I tooled around in a special golf cart speeding up slow players, stopping the odd fight and generally putting myself in harm's way. That particular course had an automatic lightning warning system that detected increased electrical fields and sounded a siren located on nearly every hole. That siren could knock squirrels out of trees, and it always astounded me that golfers could not hear it—they would just keep playing. This is what I meant by "in harm's way," because I had to put myself into dangerous situations when ordering those deaf players off the course and into shelters.

Did they appreciate my efforts? No, they went grumbling, complaining, threatening to get me fired and glancing back like angry monkeys. Golfers somehow think they're immune to terminal zap, and, every year, ambulances scream onto greens to pick up wrong thinkers. Being on golf greens during a thunderstorm is insanity, but if that should happen to you, move away from metal objects and other players, crouch down and make yourself as small a target as possible. If you're lucky, the lightning will get another golfer and not you, but hey, you can avoid the risk altogether by not playing when the weatherman warns of thunderstorms. Be smart, and live to play another day.

Floods

Great floods have flown from simple sources.

–William Shakespeare

SOGGY SITUATIONS

Backyard Waterworlds

Ontario's thousands of lakes and rivers are all subject to periodic flooding when a deluge or snow runoff overtaxes their watershed's ability to hold onto moisture. When an extended rainfall saturates a watershed, the next deluge or snow runoff will simply gush into nearby rivers or lakes, fill them beyond normal capacity and put you in harm's way. This can happen quickly, and fast-rising water can be an unpleasant surprise for local residents.

To guard against those surprises, Ontario has more than 4000 sensors at 1200 stations across the province that provide information on water levels. Ontario is well prepared. In 1992, the province amalgamated two agencies to create the Aviation, Flood and Fire Management Branch to consolidate flood, fire and aviation emergency response. If flooding is imminent anywhere in Ontario, those guys will know about it, issue flood warnings and prepare an emergency response plan. If one day that bubbling brook flowing past your house turns into a raging monster, rest assured that somebody will be there to pluck you off the roof.

The Big Kahuna of Ontario floods occurred October 14 and 15, 1954, when more than 200 millimetres of rain fell onto already-saturated Toronto-area watersheds. The Don and Humber rivers, along with the Etobicoke and Mimico creeks, overflowed their banks, ripped almost 2000 houses off their foundations, drowned 81 unfortunate people and caused millions of dollars of damage.

Spring floods followed by warm, sunny weather is a recipe for creating several billion more mosquitoes than the previous year. As the floodwaters recede, they leave behind countless standing

pools of warming water—the perfect hatching spots for mosquito larvae. In 2008, a record year of rain and flooding, entomologists reported mosquito larvae hatching rates in southern Ontario at five times normal—in northern Ontario, 10 times normal. Luckily, most flood-bred mosquitoes are your everyday nuisance variety and not the dreaded *Culex* genus that carries the West Nile virus.

Greed Isn't Good

In the past, there was little or no government oversight of Northern Ontario's mining and forest industries, and both enjoyed carte blanche in the north. Both are usually transient endeavours that require roads. When the minerals and trees are exhausted in one area, operations cease and the company moves on, but the roads remain. Trees will not set roots in hard-packed road soil, and over time, those roads can erode or slump into rivers causing floods and destroying fish habitats all the way to the Great Lakes.

Ontario's northern pines live a tenuous existence on mere centimetres of soil. Cut too many trees, and those few precious centimetres will wash into nearby rivers and cause astounding environmental damage. Huge segments of Ontario's north have become a patchwork of rocky topography unable to absorb rainwater that floods into rivers and continues the degradation of forests and fish stocks.

This legacy of greed is not confined to roads. Many of the hundreds of tailings piles sitting next to abandoned northern Ontario mines eventually slump into rivers, causing floods and spreading heavy metals over vast areas. Earthworms ingest soil containing heavy metals; small bird and mammals eat the worms and are prey to larger birds and mammals. Bear steak, anyone?

Rehabilitation of old mine-tailing sites is an expensive proposition. In the fall of 2000, the federal government undertook the rehabilitation of the Kam Cotia zinc and copper mine near Timmins. Estimated cost—$40 million, but, because of unforeseen circumstances, the project is still ongoing and cost estimates have risen to more than $55 million. What were those unforeseen circumstances? It rained more than usual during the second year of the project, turning the massive 500-hectare tailing site into a lake. The "lake" had to be contained and its acid water neutralized with lime. A messy business not easily entertained by either the federal or Ontario governments.

These days, companies cannot walk away from an unproductive mine before satisfying several government agencies that what they're leaving behind has been made environmentally safe. A good thing, but it does nothing to fix the hundreds of ecological time bombs festering in the wilds of Ontario. Meanwhile, those thousands of kilometres of access roads the mining companies hacked through the forests are still eroding and collapsing into rivers and streams, creating flood conditions and general ecological havoc. The lumber barons capped out Ontario's forest, and the miners left us hundreds of ecological time bombs—a nice legacy. Nothing to do about it now, except…

ASSUME THE WORST

Be Prepared

No matter how full the river, it still wants to grow.

–African proverb

If you live near a river or stream, be prepared for flash floods. Be alert for heavy, upstream rainfall or you could be in for a watery surprise. If you receive evacuation orders, obey them. Never attempt to cross roadways covered with water; flood currents are always stronger than they appear, and they will likely sweep you off your feet. Floods may spread sewage capable of causing a plethora of disease. If you absolutely must stomp about in floodwater, wear rubber boots and walk slowly.

It's a good idea to make a list of things that you will need and can find quickly in case of an emergency evacuation. Get adequate insurance, and if you already have a home policy, make

sure there is flood coverage. Canadian property insurers can be tricky when it comes to Act of God events, and river and stream overflow is not normal coverage in a standard policy. Here is a valuable checklist of things to do before, during and after a flood.

Before a Flood

- Make sure any photos or videos of all of your important possessions are in a safe place. These documents will help you file a full flood insurance claim.

- Store important documents and irreplaceable personal objects (such as photographs) where they won't be damaged.

- Move furniture and valuables to the upper levels of your home.

- Make sure your sump pump is working.

- Clear away debris from gutters and downspouts.

- Buy and install sump pumps with back-up power.

- Anchor fuel tanks. An unanchored tank in your basement can be torn free by floodwaters, and the broken supply line can contaminate your basement. An unanchored tank outside can be swept downstream, where it can damage other houses.

- Have a licensed electrician raise electric components (switches, sockets, circuit breakers and wiring) at least 30 centimetres above your home's projected flood elevation.

- Place the furnace and water heater on masonry blocks or concrete at least 30 centimetres above the projected flood elevation.

- If your washer and dryer are in the basement, elevate them on masonry or pressure-treated lumber at least 30 centimetres above the projected flood elevation.

- Have a family emergency plan. Post emergency telephone numbers by the phone and teach your children to dial 911. Plan and practice a flood-evacuation route with your family. Ask an out-of-province relative or friend to be the "family contact," in case your family is separated during a flood. Make sure everyone in your family knows the name, address and phone number of this contact person.

- Don't forget to have a plan for your pets.

During a Flood

- Fill bathtubs, sinks and jugs with clean water, in case water becomes contaminated.

- Listen to a battery-operated radio for the latest storm information.

- If local authorities instruct you to do so, turn off all utilities at the main power switch and close the main gas valve.

- If told to evacuate your home, do so immediately.

- If water rises inside your house before you have evacuated, retreat to the second floor, the attic or, if necessary, the roof and wait for assistance.

- Floodwaters can carry raw sewage, chemical waste and other disease-spreading substances. After contact with floodwaters, wash your hands with soap and disinfected water.

- Avoid walking through floodwaters. As little as 15 centimetres of moving water can knock you off your feet.

- Do not drive through a flooded area. If you come upon a flooded road, turn around and go another way. A car can be carried away by just 60 centimetres of floodwater.

- Electric current passes easily through water, so stay away from downed power lines and electrical wires.

- Look out for animals—especially snakes. Animals lose their homes in floods, too.

After a Flood

- If your home has suffered damage, call the agent who handles your flood insurance to file a claim. If you're unable to stay in your home, make sure to say where you can be reached.

- Take photos of any water in the house, and save damaged personal property. This will make filing your claim easier. If necessary, place these items outside the home. An insurance adjuster will need to see what has been damaged to process your claim.

- Check for structural damage before re-entering your home. Don't go inside if there is a chance of the building collapsing.

- Do not use matches, cigarette lighters or other open flames upon re-entering your property. Gas might be trapped inside. If you smell gas or hear hissing, open a window, leave quickly and call the gas company from a neighbour's home.

- Keep power off until an electrician has inspected your system for safety.

- Avoid using the toilets and the taps until you have checked for sewage and water line damage. If you suspect damage, call a plumber.

- Throw away any food, including canned goods, that has come in contact with floodwaters.

- Boil water for drinking and food preparation until local authorities declare your water supply safe.

- Salvage water-damaged books, heirlooms and photographs.

- Follow local building codes and ordinances when rebuilding. Use flood-resistant materials and techniques to protect your property from future flood damage.

Hail

*I once told Richard Nixon that the Presidency
is like being a jackass caught in a hailstorm.
You have to just stand there and take it.*

–Lyndon B. Johnson

CANNONBALLS FROM THE SKY

This Is Gonna Hurt

Imagine a tiny water droplet going for an elevator ride in a cumulonimbus cloud. The higher the elevator rises, the colder air becomes until at the top, the elevator door opens and out falls a tiny ball bearing of ice. Down goes that frozen drop into warmer air. Half melted, and almost at the ground, it collides with a friend—a tiny water drop that has never been on an elevator. Always obliging, the cumulonimbus again offers a way up, the two rise and upon reaching the upper limit, the elevator dumps out one slightly larger ice ball. The process repeats. How many times that frozen drop can fall and ride the elevator back up with a friend before finally crashing to Earth depends on the strength of the storm cloud, but it could well number in the hundreds, because hailstones can grow to a humungous size.

Another theory is similar, but instead of falling rapidly, the little drop descends ever so slowly and gathers up its little friends on the way down. Either way, the end is the same: a ball of ice hits the ground.

Hailing Records

The unofficial record weight for a hailstone is one kilogram. This monster fell with many others on April 14, 1986, in Bangladesh, killing 96 people and more than 2000 cows. There is nowhere to hide from hailstones that large and heavy—those poor folks and animals were caught in a storm of crashing cannon balls.

The heaviest officially recorded hailstone fell on Coffeyville, Kansas, on September 3, 1970, and weighed a hefty 766 grams; the largest officially measured hailstone thudded onto the ground in south-central Nebraska in June 2003. About the size of a soccer ball (17.8 centimetres), that chunk of ice got picked up and thrown into a freezer and is now preserved for posterity by the National Centre for Atmospheric Research in Boulder, Colorado.

The largest officially recorded hailstone in Canada was 290 grams (think grapefruit), and it fell on Cedoux, Saskatchewan, in August 1973. On July 14, 1953, a hailstorm pelted southern Alberta's prairie with golf ball–sized stones, carving out a swath of destruction eight kilometres wide and 225 kilometres long. The huge stones turned cows into hamburger, obliterated crops and killed an estimated 36,000 birds. Bad got even worse, when four days later, as if to add insult to injury, another hailstorm worked over the same area, killing a further 27,000 birds.

In Ontario, most hail falls onto—yes, you guessed it—the Essex county region, Ontario's tornado alley—home to big storms, serious lightning and big hailstorms. In June 2008, a hailstorm pelted Chatham with golf ball–sized stones, damaging countless vehicles and 1500 to 2000 rooftops.

On July 27, 1890, the *New York Times* reported a hailstorm had struck Embro, Ontario (12 kilometres north of Ingersoll), flattening the season's entire grain crop. The article noted that the hailstones were so large, numerous plows and wagons were required to clear the roads.

Landslides

Beware the pine-tree's withered branch!
Beware the awful avalanche!

–Henry Wadsworth Longfellow

SLIP SLIDING AWAY

The Earth Moves

For most residents of Ontario, landslides are the least worrisome of natural calamities. It has never happened before, why should it happen now or in the future? Not so with residents living along the St. Lawrence and Ottawa River Valleys, as those folks live in constant fear of slip 'n' slide. The stuff under their feet is Leda clay, a soil composed of glacier-ground rock dust and silt. Some 10,000 to 12,000 years ago, it layered the bottom of the Champlain Sea, a huge body of water that rushed in from the Atlantic to fill depressions left by retreating glaciers. During the next 3000 years, that sea drained out through the St. Lawrence River, leaving behind land that's more water than soil and prone to slippage when deluged by rain.

The soil's unstable character is caused by the loss, by rainfall, of its original clump-binding salt. This left the land susceptible to slumping, with no ability to reform. Once started, the soil simply runs off like thick soup, plugs rivers and causes flooding. Scientists have identified more than 250 major landslides that occurred within a 60-kilometre radius of Ottawa during the last 7000 years. Called "quick-clay landslides," they can be initiated by deluge, snowmelt, earthquakes or construction projects. During the last century, there have been at least 23 major slides in an area between Ottawa and Québec City, with more than 100 fatalities. In this region, land can be stimulated to move by the simple act of forgetting to turn off a lawn sprinkler.

Ontario experiences one major landslide every 10 years and hundreds of minor slides annually—the majority involving falling rocks and subsequent road closures in Northern Ontario. Municipalities play a guessing game trying to mitigate rockslides, and every year a different road is closed. Not

LANDSLIDES

much to do about that, but there is something Ontario can do about the building of structures on unsafe land sites. Most municipalities in Ontario don't allow the construction of structures on flood plain, but how is it that people can build on unsafe rocky slopes and Leda clay?

In 1971, in response to a huge flow slide on the Saguenay River near St. Jean-Vianney, Québec, that claimed 31 lives, and another on the South River a few days later, the South Nation Conservation Authority tested soil the entire length of the river. At the small town of Lemieux, they found Leda clay and proclaimed the area unsafe for habitation.

The Ontario government did nothing for 20 years, until a small slide in 1992 caused the South Nation Conservation Authority to forcefully demand that the Ontario government wake up and move the town to a safer spot. Better late than never, because had the government waited another year, bureaucratic heads would have rolled. On June 30, 1993, the abandoned town of Lemieux suddenly dropped 15 metres and slid into the South Nation River.

LANDSLIDES

Landslides occur frequently on the Great Lakes—Erie and Ontario especially—with shore boundaries composed of different strata of glacier clay. Bluffs are common along these lakeshores and prone to collapse. Nowhere is that more apparent than at the Scarborough Bluffs just east of Toronto, where their origin can be read like a book. Kids playing on the cliff-tops today had grandparents who build cottages there; both practices are now discouraged and for good reasons. On August 14, 1994, a high bluff near Port Burwell on Lake Erie collapsed, burying four children who narrowly escaped suffocation—and a good number of those cliff-top cottages have slipped into the abyss.

Smog

*...it does our lungs and spirits choke,
Our hanging spoil, and rust our iron.*

–from the ancient "Ballad of Gresham College"

OUT WITH THE BAD AIR

Breathing Can be Hazardous to Your Health

Every year, during the warmer months, Ontario's air turns poisonous and, according to the Ontario Medical Association, kills almost 10,000 residents prematurely. It primarily affects the very old and the very young, because both groups are particularly susceptible to that nasty mix of airborne particulates and low-level ozone called smog.

In the old days, cities of a newly industrialized Europe burned soft coal to produce heat and were constantly enveloped in a smoky, green fog that tasted of matches. Over the years, the green fog became worse, and residents suffered. But none suffered like the people of London, England, where smog was soon the norm and became known as "pea soup." Nasty stuff, but on one chill winter day, nasty became a disaster.

SMOG

On December 5, 1952, a cold fog descended on London, and residents lit up a million or so coal fires. Trapped by a cold-air temperature inversion, smoke from those fires piled up to such an extent that the elderly, newborns and people with respiratory problems began to die. It lasted five days and killed 4000 citizens, with 8000 more dying during the following weeks. A major catastrophe, but it led to a good thing—the passing of Britain's 1956 Clean Air Act and the outlawing of coal-burning fireplaces.

So a good thing came from bad, but smoke was not responsible for killing all those people. Another devil lurking in the green haze was low-level ozone. London's residents couldn't see it, but above the smog, sunshine changed the hydrocarbons and nitrous oxide in the smoke to ozone. Exposure to ozone causes respiratory problems and can aggravate any number of existing health issues. People supply most of the raw materials needed to manufacture ozone through industrial pollution or vehicle emissions; the sun simply finishes the job. Most is the key word here, because nature also contributes a good deal of raw material through volcanic eruptions, forest fires and tree emissions.

Ontario gets most of its weather from the U.S., and there are days when provincial residents can taste Chicago, Pittsburgh and other nearby American cities. In addition, Ontario residents suffer from nitrous oxide emissions emitted by forests; airborne particulates and sundry chemicals from rampant forest fires; crappy effluent pouring out of six badly scrubbed coal-fired electric plants; and more vehicles on the roads. It's not getting any better as more people arrive to crowd Ontario's already overpopulated cities and motorways. Seems like a no-win situation, but it's still better than Québec, which is downwind and gets the crappy air after Ontario is done with it…cough, cough.

Pine Smog

Ontario is pine-tree heaven. We have billions, and every one is pumping volatile chemicals into the air we breathe. Waxes, nitrous oxides and terpenes all smell nice and healthy, but, when airborne and in contact with sunshine and ozone pollution, they form aerosols (microscopic solids suspended in air) similar to industrial and automobile emissions. Water molecules are attracted to aerosols and can form snowflakes or rain drops. Nice, because rain and snow cleanse the air and make it well again. Not nice are companies that clear huge tracts of forest for large-scale lumbering and farming operations.

Large-scale planting of pine trees to act as "carbon sinks" has been promoted by scientists all over the world, the theory being that fast-growing trees absorb and retain carbon dioxide from the air. Carbon dioxide is widely cited as one of the major contributors to global warming. Planting pines was considered a good thing, until researchers in Australia discovered pine trees were producing nitrogen oxide. Nitrogen oxides are smog precursors; they combine with other pollutants to form ground-level ozone, a major component of smog.

The researchers found that, although the amounts produced were insignificant on a local scale, global nitrogen oxide issuing from boreal coniferous forests might be comparable to those produced by worldwide industrial and traffic sources. You will not hear a word about pine smog from the green-for-bucks media crowd, because it rubs their story the wrong way, but it's out there and coming from Ontario's forests at a rate comparable to having two vehicles in every garage.

Forest Fires and Insects

In elementary school, you were told that in case of fire you had to line up quietly in a single-file line from smallest to tallest. What is the logic in that? Do tall people burn slower?

–Warren Hutcherson, comedian

BURNS AND BUGS

Towering Infernos

What does weather, weird or otherwise, have to do with forest fires? Answer is…plenty. Weather is the cause of most forest fires, and if that weather is only slightly weird, the resulting conflagrations can be catastrophic. Inversely, weather can have a damping affect on forest fires, as during the 2008 fire season (April 1 to October 31), when northern Ontario's forests were extraordinarily wet because of heavy rainfall, and forest fires were at record low numbers.

Smokey the Bear and his famous saying: "Only you can prevent forest fires" was promotional hooey. Although careless campers and smokers account for a small percentage, lightning is the prime instigator of Ontario's forest fires. Millions of hectares of forest are torched by lightning annually, and in some years, the affected areas are larger than many countries. Forest fires have cost the province of Ontario billions of dollars and do major damage to the air we breathe.

To counter this, the Ontario government has set up a province-wide system of lightning detectors. Threatening storms are tracked, lightning strikes counted and fire-fighting teams put on the alert. Strikes can number 100,000 to 150,000 per week, and when a fire or fires break out anywhere in the province, teams are into the air and on the spot within hours.

But, sometimes, being Johnny-on-the-spot does not help. During August 2005, northern Ontario suffered more than 130 simultaneous fires. (The photo shows smoke plumes from numerous fires.) The largest, a monster called Thunder Bay 57 for the closest city, became so huge that firefighters gave up the fight and left more than 5000 hectares of forest to burn.

Abandonment is not always an option; in populated areas, firefighters must dig in and battle the beast to a standstill. To do this, Ontario relies on neighbouring provinces and U.S. states, because sharing resources can make an army of fighters and fleets of water bombers available to battle fires before they "escape." Escape is a term used by experts to describe a forest conflagration not immediately addressed by suppression teams—a missed opportunity. Escaped fires do significant damage to Ontario's forests, and to counter these, the government employs a vast network of fire spotters working from towers and aircraft.

During the summer months, a red sun at dusk usually indicates a fire is raging somewhere in the west, and you're probably already breathing its noxious output of poison gas. During the summer of 2004, 12 percent of the Yukon and much of Alaska were on fire, sending an estimated 30 billion kilograms of carbon monoxide and carbon dioxide into the air, along with heavy metals such as mercury and nickel.

Canada suffers an average of 10,000 forest fires and loses roughly two million hectares of forest annually, an area larger than most U.S. states. Emissions from these fires are equivalent to about 25 percent of Canada's annual industrial and vehicle pollution output; an astounding amount, and, with the country in a warming phase, that will only get worse.

More fires, more pollution and a federal government loathe to increase funding to fight fires will mean higher health costs and increased taxes for Ontario, a province with an enviable record of fire suppression. Fires burn an average of 100,000 hectares of forest in Ontario, a modest amount compared to the national loss of two million.

Insect Devastation

Provincial fire losses pale by comparison to the damage done by invasive insect species brought into Canada via packing crates. These critters munch a million hectares of Ontario forest every year. The province will be unrecognizable in 25 years: no spruce, no white or red pine, no majestic oaks, maples, ash, hickory or walnut trees. Ontario's forests might go the way of the Atlantic salmon—replaced by orderly stocks of genetically altered and farmable trees that are good for making biomass ethanol.

The transformation is probably inevitable, because Ontario's ability to control insects is countered by its insatiable appetite for cheap overseas trade goods—a new species of bug seems to come in every crate. According to a study published in the science journal *Frontiers in Ecology and the Environment,* alien-species invasion ranks just behind habitat destruction as the greatest cause of species extinction.

What goes around comes around, and this cheap-trade-goods fiasco might be a fitting revenge for Native peoples—except they're in the same pickle. Joni Mitchell's famous song, "Big Yellow Taxi," has a prophetic line that goes:

> *They took all the trees,*
> *put 'em in a tree museum*
> *And they charged the people*
> *a dollar and a half just to see 'em.*

Somewhere down the road, Ontario's forests will go the tree-museum route, but most will look the same as today's, except the trees will be smaller and of a uniform size. Northern Ontario will resemble a gigantic Christmas tree farm, and the south will look like everyone's backyard. If you doubt that, ask your grandfather about the forests of his day—about the tall trees and how he would spend a whole day climbing to their tops. Then again, he might recall talking to his granddad about the skyscraper forests that got turned into tall masts for sailing ships.

Winter's Nasty Bite

Mon pays, ce n'est pas un pays, c'est l'hiver.
(My country is not a country, it's winter.)

–Gilles Vigneault, singer

WINTER'S NASTY BITE

BLIZZARDS AND COLD

Baby, It's Cold Outside

Environment Canada officially classifies a blizzard as having these components: snow or blowing snow with winds of 40 kilometres per hour or more, visibility reduced to one kilometre, a wind chill of –25°C or colder, and lasting four hours or more. A conservative yardstick for a weather phenomenon capable of putting the brakes on all our worldly activities.

There are two types of blizzards: a regular precipitation blizzard and ground blizzards, where the precipitation is already present and snow is moved by strong winds. Ground blizzards

are divided into three categories: horizontal advection, when the wind blows across the surface; vertical advection, when the wind lofts snow upward, creating large, drifting waves; and thermal-mechanical, when both wind and updrafts create massive rolling waves called snow billows. Billowing occurs mostly in arctic areas and is extremely dangerous, because it makes vision almost impossible and breathing difficult. Ground blizzards occur mostly on vast, open areas without trees to catch snow.

Open areas, such as Lakes Erie and Ontario, are more susceptible. In the weeks before the new year of 1977, a shifting arctic air mass caused migrating warm air from the U.S. Midwest to dump massive amounts of snow onto the frozen lakes; because of the cold, it stayed dry, fluffy and ready to shuffle off to Buffalo. On January 27, 1977, a cascading arctic cold front slammed onto frozen Lake Erie, and all that dry snow shuffled off to become the infamous Blizzard of '77, a storm that caused major problems for southern Ontario's Niagara region…and Buffalo, New York.

Winter cold is no joke; it kills more Canadians than all other weather phenomena combined. In Ontario, cold weather is guaranteed in the north, and residents are prepared for eventualities, but in the south, cold weather is variable, and most folks are unprepared. On Christmas Day, 1980, residents of Toronto endured –25°C temperatures—on the same day in 1982, the Toronto temperature was a balmy 17°C.

Winter causes driving conditions to deteriorate, and southern Ontario residents suffer from variable weather attitudes. During the 2009 winter driving season, the OPP investigated more than 17,000 weather-related vehicle accidents, many involving fatalities. Residents of southern Ontario will drive regardless of the road conditions; staying home in the worst winter weather is not a consideration, and driving at normal speed and talking on a cell phone are considered a commuter's right. A dangerous situation, only compounded by the sheer number of immigrant drivers unaccustomed to winter driving. Southern Ontario's highways are death traps in winter, and, although the accident might be no fault of yours, dead is dead.

If you love anything—life, family or a pet—do them and yourself a favour and stay home when winter weather threatens to turn weird or just plain bad. If this is unavoidable, drive defensively and go prepared, because that few kilometres from the highway to your house could become endless if you should skid off a deserted road. If you must travel in winter, have a survival kit in your vehicle.

A Driving Survival Kit
(courtesy of Environment Canada)

Antifreeze	First-aid kit
Axe or hatchet	Flashlight and new batteries
Blanket	Ice scraper and brush
Booster cables	Salt
Candle and matches	Sand
Compass	Shovel
Emergency food pack	Tow chain
Extra clothing and footwear	Warning light or road flares
Fire extinguisher	

WINTER'S NASTY BITE

FROSTBITE AND HYPOTHERMIA

Frostbite

Here are some useful tips to help prevent and (if necessary) deal with frostbite, courtesy of the Canadian Red Cross:

- Wear a hat and clothing made of tightly woven fibres, such as wool, which trap warm air against your body. A few light layers protect better than one heavy garment. Protect vulnerable areas such as fingers, toes, ears and nose.

- Drink plenty of warm fluids to help the body maintain its temperature. If hot drinks are not available, drink plenty of plain water. Avoid caffeine and alcohol, which hinder the body's heat-producing mechanisms and will actually cause the body's core temperature to drop.

- Take frequent breaks from the cold to let your body warm up.

- Signs and symptoms of frostbite include:
Numbness
Tingling
Pain and swelling
A total loss of sensation
Pale waxy skin becomes dark blue in colour; in severe cases, skin will look charred and burnt.

- In case of frostbite, cover the affected area. Never rub the skin, because this can cause further damage. Instead, warm the area gently by immersing the affected part in warm water. Bandage the affected area with a dry sterile dressing. Ensure that the affected part does not become frozen again, and get the individual to a doctor as soon as possible.

Hypothermia

And now more useful tips, this time to help prevent and (if necessary) deal with hypothermia, again courtesy of the Canadian Red Cross:

The signs and symptoms of hypothermia include:
- Feeling cold
- Shivering (which will stop as the condition worsens)
- Slurred speech
- Pale skin, bluish lips
- Slow pulse
- Mood swings
- An inability to think clearly
- Lethargy and ultimately unconsciousness.

- In case of hypothermia, start by removing wet or cold clothing, and replace with warm, dry clothing. Keep the individual warm by wrapping him or her in blankets and moving the victim to a warm place. Remember to be gentle when handling the person. Never rub the surface of the body; this could cause further damage if they're also suffering from frostbite.

- If the individual is dry, use hot water bottles or heating pads to warm them. Make sure there is a blanket, clothing or towel between the heat source and the person's skin. If the person is awake, give warm liquids to drink. Avoid alcohol and caffeine, because they can hinder the body's heat-producing mechanisms. Do not warm the individual too quickly by immersing him or er in warm water. Rapid re-warming can cause heart problems.

WINTER'S NASTY BITE

GIANT SNOWFLAKES

They're How Big?

*How full of the creative genius is
the air in which these are generated!*

*I should hardly admire more if real stars
fell and lodged on my coat.*

—Henry David Thoreau

Think falling snow and our minds might conjure up an image of Christmas Eve and blurry streetlights. Skiers might visualize fun in the powder, police officers slippery roads and hard work ahead, and pilots a wall of white that impairs vision on landings and takeoffs. Different situations, but the falling snow visual is universal—many indistinguishable flakes falling to the ground. It's like the weather: expected.

How do you think your mind would respond to the unexpected—snowflakes big as pie plates, for example? Snowflakes are super-cooled water droplets held aloft by upward-sloping moist winds that allow the droplets time to bind with other drops and crystallize. As the six-armed crystals grow, they become heavier, until they fall to Earth as intricately formed flakes. Sound familiar? It should. Hailstones form the same way, except that, with snow, the water drops have had time to crystallize. The process that makes large hail will also produce huge, "blob" snowflakes of the more-common slushy variety.

Huge crystallized snowflakes are extremely fragile and break in the wind, which gives reason to their rarity. Occasionally, depending on conditions, you might be lucky enough to see a sky filled with gently falling, pie-tin-sized flakes. The largest snowflakes, according to an article in a 1915 issue of *The Monthly*, fell on Fort Keogh, Montana, on January 28, 1887, and measured 38 centimetres across, larger than a big pie plate. In 1951, giant snowflakes fell onto the English town of Berkhamsted, and in 1971, huge 30-centimetre flakes were recorded falling in Siberia. William S. Pike, a British weather observer for the Royal Meteorological Society, has searched old records and found almost a dozen reports of huge flakes. He claims that, during winters, the maritime fringes of North America and Asia probably see a fall of giant flakes almost every day, but they just aren't reported.

Giant flakes are out there, and finding them has become a challenge for NASA. In 2013, they'll launch a satellite to find and measure giant flakes. In doing the research on giant flakes, I found dozens of reports, and some of those came from Ontario's near-Arctic settlements. Go north, young man (or woman), and look around for wonders.

Weather Modification

*The significant problems we face cannot be solved
at the same level of thinking we were at
when we created them.*

–Albert Einstein

WEATHER WIZARDS

Rainmakers

People have been tinkering with the weather for years. During the 19th and early 20th centuries, travelling medicine men crisscrossed rural America and Canada, putting on shows, hawking patent medicines and trying their hand at rainmaking. They were not very successful—until the newspapers discovered James Pollard Espy.

In 1841, Espy wrote a book, *The Philosophy of Storms,* in which he laid out in print his discoveries and observations of the mechanics of thunderstorms and how stubborn weather systems could be induced to precipitate by introducing soot particulates into updrafts. Espy cited the observations of

Benjamin Matthias of Philadelphia, who, on repeated visits to English industrial cities, noticed that it rained every day except for Sundays, when the factories shut down. Espy theorized that carbon particulates from industrial pollution acted as a catalyst to promote rainfall.

Espy's book became a hit with those imaginative travelling medicine men, and they immediately began using various and sundry polluting devices to seed clouds. The newspapers of the day loved James Pollard Espy and called him the father of modern American meteorology. Pure sensationalism, but it lent credence to rainmaking, and it became a semi-respected occupation. Neat, but nobody got rich. At least, not until 1848, when something happened to change the fortunes of rainmakers; gold was discovered in California, and thousands of opportunistic folks moved westward.

What does gold in California have to do with Canada? A good percentage of folks stampeding to California were Canadians, and many went via the Canadian prairie before turning south. Those opportunistic rainmakers followed and discovered something interesting along the way.

Smudge-pot Science

Every year, the Native peoples on the plains torched huge tracts of prairie grass to promote new growth and attract buffalo. Burned black one year meant green and lush the next and plenty of buffalo. It also meant settlers had to wait out the dangers of prairie fires, and they waited mostly in pouring rain and smoke—a weird weather phenomenon not lost on the travelling medicine men.

Lots of smoke (which sent soot or carbon particles into the air) was the key to forcing stubborn rain clouds to give up their moisture, so those travelling rainmakers invented the "smudge pot." They filled huge iron pots with rock oil (crude oil) and burlap, placed a dozen or so in a circle and set them ablaze. Impressive, but the fact that they worked some of the time was even more impressive. These snake-oil salesmen became the first weather-makers. A few got so good at wringing rain from clouds, they became famous and wealthy men.

In the early 20th century, rainmakers found Canada a lucrative stomping ground—the railway was complete, and World War I had tripled the price of wheat. Large farms sprang up everywhere; however, Canadian farmers would have no part of burning prairie land, forcing the rainmakers to use more and more smudge pots. Smudge-pot rainmaking died out completely before World War II, though smudge pots are still in use today, having been adapted to protect citrus crops, because smoke particles and heat prevent the formation of frost.

Weather Modification Works, But...

After World War II, rainmaking enjoyed a rebirth, when two American researchers, Vincent Schaefer and Bernard Vonnegut (author Kurt Vonnegut's smarter brother) discovered that they could make rain by using dry ice and silver iodide crystals to "seed" clouds from airplanes. The two ingredients act as nuclei and attract water droplets until they reach a size large enough to precipitate from the cloud as rain.

The chemicals for seeding clouds can be introduced by airplanes or from the ground using pyrotechnics, chemical generators, artillery and rockets. China is a big believer in weather modification and spends about $100 million annually on projects that employ roughly 30,000 scientists to do research and experiments. The Chinese use more than 7000 artillery guns and 4000 rocket launchers in cloud-seeding efforts nationwide; they have enjoyed some success and caused a few problems.

One problem occurred in 2007, while the Chinese were engaged in pre-Olympic practice to mitigate weather. China's English-language newspaper, *The China Daily*, reported that three government planes carrying 30 technicians had flown for three hours over Inner Mongolia, dropping silver iodide and small amounts of diatomaceous earth (the skeletal remains of tiny, ancient sea creatures composed mainly of silica) into the atmosphere at about 8000 metres. Some hours later, Beijing experienced a period of extraordinary thunderstorms and suffered extensive damage. Even though the news outlet was politically unable to point fingers, the article left readers to make the connection.

Weather modification works, but there are inherent problems, control being only one. Did seeding clouds in Mongolia cause damaging storms in Beijing? Perhaps, but it can also work the opposite way. Inducing rainfall in Mongolia could deprive Beijing-area farmers of precipitation. Dangerous stuff, but presently more than 150 weather-modification projects are running in more than 140 countries.

During the early '60s, a rainmaking project called Operation Umbrella began in the Lac Saint-Jean area of Québec, a region

that was experiencing long-term drought conditions. The sun always shone in Lac Saint-Jean, but during the few years of Operation Umbrella, it rained almost every day. It rained so much that concerned mothers in the area petitioned the Québec government to supply them with vitamin D to protect their children from rickets, a disease caused by lack of sunshine.

Newspapers picked up the story and vilified the Québec government for raining down chemicals on helpless children. Horrified by the possible political repercussions, the Québec Minister of Natural Resources ordered a halt to all rainmaking activities in that province—a ministerial edict that had far-reaching effects across the country, since political timidity is not confined to Québec.

WEATHER MODIFICATION

HAIL SUPPRESSION

Please Make it Stop

Canada is currently funding a hail-modification project in Alberta, but it's a bit Johnny-come-lately—a group of Canadian insurance companies calling themselves the Alberta Severe Weather Management Society (ASWMS) has been running the Alberta Hail Suppression Project (AHSP) for almost two decades. The success of this project is dramatic and responsible for a 50-percent reduction in hail insurance rates for Alberta farmers and ranchers, an annual savings of almost $50 million per year—a big deal for hard-pressed grain farmers.

Hail is most common in mid-latitude areas, such as Alberta, during summer, when surface temperatures are warm enough to create the water vapour and convective activity necessary for thunderstorms, but the higher atmosphere is still cool enough to support the formation of ice. Evaporated surface water is carried aloft by warm, rising air and reaches cooler, higher altitudes, condensing into water droplets. Caught in strong updrafts, these droplets rise above the freezing level and become supercooled. During the summer months, dust is more prevalent and is sucked up to high, cold altitudes by updrafting altocumulus clouds, where it serves as hail-forming nuclei and attracts supercooled water droplets. To suppress hail, aircraft or rockets introduce a large quantity of rain-making crystals to attract water droplets, thereby making the moisture unavailable higher up. This mechanically attracted water vapour falls to the ground as rain—an undemocratic process if you happen to be the deluged farmer.

Alberta is Canada's Hail Alley; damages to crops by hail average more than $100 million a year and can sometimes be much more. During the summer of 1991, a single hailstorm caused almost a half billion dollars in crop damage. Little wonder Alberta is on the forefront of weather modification.

Ontario's Hail Alley, the Essex County area, experiences hailstorms almost every year, and a few have been very expensive. On June 9, 2008, a hailstorm ripped through the area, flattening crops, denting cars with golf-ball-sized missiles and causing millions in damages. Two weeks later, another hailstorm rampaged through the area, doing exactly the same thing.

During the 2008 Ontario apple-growing season, hail insurance claims by farmers exceeded $8 million—little wonder farmers (and insurance companies) are clamouring for weather-modification programs like those in Alberta. Problem is, tinkering with weather in one spot has it tinkering with you in another. A weather system prodded to release precipitation in expectation of hail might well deprive other areas of needed moisture. This, in turn, could drive small farmers out of business (or force them to sell to large agricultural conglomerates), and small farms are the backbone of Ontario's agricultural sector. Weather modification works, but who will benefit is a big problem for Ontario, a province still studying the issue, while conducting small experiments in the north.

Atmospheric Optics

*The older I get, the surer I am
that I'm not running the show.*

–Leonard Cohen

LOOK UP, OR MISS THE SHOW

It's a Bird. It's a Plane. It's…

Ontario is an atmospheric battleground, a place where great armies of molecules attack and retreat, expending huge amounts of energy in the process. As with any conflict, there are times when the combatants become drained of energy and a truce is declared. Skies clear, the sun or moon shines brightly and all is peace and tranquility. For sky watchers, this is half time and, as at a football game, they expect a good half-time show—an atmospheric optical extravaganza. Rainbows are the usual offering, but nature has a huge repertoire and I have listed only a few. To see more of these atmospheric treats with pictures go online; there are dozens of sites worth exploring. You might even become a sky watcher and start recording your finds with a camera.

Moonbows

Rainbows occur when water droplets refract sunshine. If the moon is full and low in the sky, it can treat you to the same phenomena, but it's called a moonbow. They're not as bright as rainbows, usually appear white and are so ethereal you can actually believe there is a pot of gold at the end. Moonbows are not that common (in addition to the full moon, you need rain happening opposite the moon), but when conditions are right, you'll see one and think you've found a pot of gold.

Sun Dogs

Three suns? You might see them and question your sanity. But you're all right, they're just bright segments of halos. Like rainbows, sun dogs are caused by refracted sunshine, but instead of water droplets, the refractors are high-altitude, hexagonal plate-shaped ice crystals aligned horizontally to the sun. Most sun dogs exhibit a spectrum of colours, from red (closest to the sun) to a pale bluish white fading away from the solar orb. These dogs are at their brightest when the sun is low and are most often seen during the winter months.

A few sun dog events have become famous. In Nuremberg, Germany, on April 14, 1561, thousands of people saw a sunrise sky filled with celestial wonders. Neat, and the artist Hans Glaser portrayed the objects in a 1566 woodcut called the *1561 Nuremberg Event*. UFO believers think the woodcut depicts a clash of spacecraft, but the phenomena Glaser depicted are likely parhelia (sun dogs), halos and floating ice crystals called diamond dust.

Circular Halos

Rainbow-like halos that surround the sun are not rare; the reason they're not noticed is that, these days, nobody bothers to look up anymore. A solar halo is a collection of millions of ice crystals, floating five to eight kilometres up, that happen to be perfectly aligned to refract sunlight right into your eyeball. The 22-degree halo around the sun is most common; it gets its name because its inner edge is 22 degrees from the sun. Like sun dogs, halos show a rainbow of colour, with red on the interior. Sometimes you'll get lucky and see a twin bill: a 22-degree halo with sun dogs attached. Halos aren't just a daytime event. Keep an eye out for one when the moon is full, though lunar halos are quite dim and colourless.

Diamond Dust

Ice crystals are most obvious when forming halos around the sun or moon. But they can also appear as precipitation and make you a believer in enchanted lands and fairies. These ice crystals form in high cirrus clouds and tumble to earth through clear skies reflecting sunshine like tiny diamonds. It is an enchanting phenomenon to witness, especially when the air is still and the crystals hang in the air nearly motionless. When this occurs, it turns adults into small children and children into the happiest beings on the planet. You'll see diamond dust mostly in the Arctic, and to experience both diamond dust and the northern lights will have you flying the flag and not wanting to leave. Until you realize it's −20°C or colder and time to get inside.

Sun Pillars

These can be startling phenomena and have you thinking UFO. The sun is rising or setting and suddenly there is a giant beam of bright light rising from the ground. Relax. What you're seeing is sunlight reflecting off millions of plate-like ice crystals tilting on their vertical axes. Sun pillars are neat and can sometimes extend to 20 degrees above the horizon. Although mostly white, they can sometimes treat audiences to a display of vivid colours. They will often hang around for an hour or so after sunset (or appear well before sunrise). The formation of a sun pillar can be compared with the glitter path caused by the setting sun reflecting on a wavy water surface. Look for them; they're not rare and occur about 30 times a year in most places.

Fire Rainbows

These resemble a rainbow set aflame and appear when hexagonal ice crystals, suspended in a high-altitude cirrus cloud, are properly aligned with the sun. To create this fiery sight, sunlight must enter through a side facet of the ice crystals and exit though the horizontal bottom face. If conditions are perfect, light refraction will transform the entire cloud into a flaming rainbow. A fire rainbow, also known as circumhorizontal arc, is a not-to-be-missed optical display, but, alas, they're so rare you'll probably never see one except in a photograph. National Geographic has a good one on its website.

Noctilucent Clouds

The sun has set, no moon is present, but there is an amazing light in the sky. Sometimes coloured, sometimes moving, a noctilucent cloud can look intriguingly like a UFO. If you see one, relax; it's just the sun reflecting off an extremely high cloud (80 to 100 kilometres high) that's probably composed of ice crystals sticking to both meteor and volcanic dust. Civil authorities hate these clouds because reports of UFO sightings pour into their offices after every appearance. That aside, anyone who spots one of these clouds is thoroughly impressed, because they're a magnificent phenomenon.

First sighted by sunset watchers in 1885, two years after the massive eruption of a volcano called Krakatoa, the clouds persisted after the volcano-inspired sunsets had stopped. Astronauts see noctilucent clouds from outer space and often wax poetic about the wondrous, electric-blue clouds that waft in the fringes of Earth's atmosphere. Why these clouds are in the mesosphere is a mystery, because no water vapour exists in that sphere to produce ice crystals. Theories as to their origin abound, but most centre on volcanic and meteor dust gathering up traces of induced water vapour like tiny mops—water vapour probably introduced into the mesosphere via volcanic eruptions and rocket contrails.

Nacreous Clouds

These are super-high stratospheric clouds, at altitudes of 15 to 25 kilometres, that might contain water mixed with nitric and sulphuric acids—the stuff that destroys ozone molecules. These clouds are rare and appear as a band of vivid pastel colours. Both noctilucent and nacreous clouds are fast movers and have been recorded travelling in excess of 750 kilometres per hour.

Sometimes called mother-of-pearl, nacreous clouds are mostly visible before dawn or after dusk, when they catch sunlight from below the horizon and blaze with bright and slowly shifting iridescent colours. They're gossamer sheets of cloud that slowly furl and unfurl in a dark sky, putting on a spectacular show.

Crepuscular Rays

Sunbursts—sometimes called the Fingers of God or Jacob's Ladder—are optical effects caused by sunlight streaming through holes in clouds and made visible by particulates or water vapour in the atmosphere. Crepuscular rays appear to converge on the sun, and if they converge in the opposite direction, they're called anti-crepuscular rays. Like crepuscular rays, these are parallel shafts of sunlight from holes in the clouds, and their opposing direction is simple perspective. For example, a wide road converges toward the horizon, but turn around and it converges to the opposite horizon.

Green Flash

Vacationers look for it religiously—the legendary emerald-coloured flash of light that occurs seconds before sunrise or sunset. Some people swear it doesn't exist, but don't believe them. It's there all right, but only for a second—so don't blink. A green flash occurs because sunshine is refracted by our atmosphere and curves ever so slightly. Green and blue light bends more than red and orange light, so green/blue rays from the upper limb of the setting sun are the last light we see as the solar disk vanishes (or appears if you're watching at sunrise).

The horizon must be clear of clouds to see the flash, and in the humid tropics, where most people watch for it, the green flash is a rare atmospheric condition. Of course, this explains why some folks swear it doesn't exist. If you're adamant about seeing the green flash, make the Sahara or Mojave desert your next vacation destination, or spend every evening on your next cruise watching the sunset.

Airglow

Airglow is why night is never absolute darkness. Our atmosphere creates light—not much but enough to see your hand in front of your face. It does this in various ways: the recombining of ions split apart by sunlight; cosmic rays entering the exosphere; or by simple chemical luminescence, as when atoms of oxygen and nitrogen combine to form a molecule of nitrous oxide and emit a photon. Airglow enabled our early settlers to find their outhouses in the dark and was considered another of God's gifts to man. But modern astronomers, forced to send their telescopes into space to avoid airglow, have a different slant and consider it an expensive curse from somewhere else.

Aurora Borealis

Northern lights to Canadians, aurora australis to those down under and a great show in the polar reaches of either hemisphere—the dance of the wispy veils that people will sit and watch for hours. Green is the most common colour, but red, blue and violet have occasionally been observed. Green arcs are most common, but during a good display, rays and curtains of various colours will swish and swoop across the sky. You can find great websites featuring aurora borealis photographs on the Internet.

These shimmering lights are caused by particles from the sun colliding with gas molecules in our thermosphere, sometimes called the ionosphere. When solar particles collide with those atmospheric gases, the collision energy between the solar particle and the gas molecule emits as a photon—a light particle. When there are millions of collisions, you have an aurora—lights that move across the sky.

Northern lights occur in southern areas of Canada about three times a year; in the far north six times or more per month. The best viewing seasons are early fall and early spring, but if you're really intent on seeing a show, hop on a plane to Iceland or northern Norway, where they dance almost every night.

Ontario Weather Stories

The one who tells the stories rules the world.

–Hopi proverb

HOT TIMES IN TIMMINS

"Say cheese," joked the news photographer and got laughs from the two Ontario Provincial Police (OPP) constables. The photographer laughed too, but thoughts of the pair embarking on a three-week dog-sled trip through snow and dense bush all the way to Moosonee soon had him shaking his head. Who would be crazy enough to do something like that? He thought for a while, and, with pen poised, asked that question of Constable Erik Howells of Thunder Bay and Constable Guy Higgott of Orillia.

"We just wanted to do something to commemorate some of the hardships our fellow officers went through back in the old days," Higgott said. "It was a tough job, but somebody had to do it."

"Can you say minus forty degrees?" said Howells. "All we're taking is a small wood stove and a tent. But we're certainly dressed warmly."

Their re-created woolly uniforms were not standard issue until the 1920s, but on a trip to commemorate the 100th anniversary of the founding of the OPP, it's what a guy wears, along with snowshoes.

The woolly duo met 10 years earlier and discovered a mutual love of the outdoors. It was an affinity they parlayed into their police activities: mountain warfare training in Vermont, extended trips into the bush and a mountain climbing expedition in Bolivia, where they dreamt up the commemorative trip.

"We wanted to try and put something together that would honour the hardships and resourcefulness of the officers that have gone before us and created the foundation (for) what the

OPP is today," said 14-year veteran Higgott, who's based at the OPP Academy in Orillia, Ontario.

On their return from Bolivia, the duo pitched the idea to their bosses in Orillia. They thought it a good idea, as did the OPP museum staff in Orillia. Given the green light for a commemorative trip, Higgott, Howells and the museum staff delved into the historical background of such a venture.

"It's a unique opportunity for both us—although in a much easier environment—to walk in the boots of the officers who've gone before us," Higgott said. "Any hardships we will face on the trail are probably minuscule to what a lone OPP officer up in the frontier north was policing. It was definitely a step back in time for us."

On February 1, 2009, the pair mushed out from Hearst accompanied by a guide to point the way to Kapuskasing. Then it was on to Timmins, South Porcupine and Cochrane—with a nine-day mush to Moosonee. On their first night under the stars, they learned that dogs howl at the moon to keep warm in −40°C temperatures, and, with three teams trying to keep warm, grabbing a few winks was difficult.

During stops along the way, the pair gave survival and bush skill demonstrations to locals. After arriving at Moosonee, they headed back to Cochrane by train on February 20. Aside from the cold, the pair mushed through fair weather all the way and never once had to contemplate eating their dogs.

To actually step back in time to, say, the year of the OPP's creation in 1909 would probably have Howells and Higgott thinking differently about mushing 1000 miles to Moosonee or anywhere else in the north. Hearst, in 1909, the starting point of their trip, was a four-shack whistle stop for the uncompleted National Transcontinental Railway (NTR) and was then called Grant. Kapuskasing, their first stop on the way to Moosonee, was called MacPherson and consisted of a few shacks and a water tower built by the NTR.

Built by the federal government, the NTR would run east from Winnipeg to Moncton when finally completed in 1915. Newly minted Ontario Provincial Police constables would not have been welcome in Hearst, Kapuskasing, Cochrane or any other railroad whistle stop. Railroads had their own police forces that answered only to Ottawa. Further north, the constables would have had to contend with Hudson's Bay Company factors. Rupert's Land had been purchased by Canada in 1868, but the HBC still ran the show, and the Company brooked no interference in jurisdictions.

Ontario mining and logging companies had followed the railroad north like piranhas, and, once established, they behaved like kings. The logging companies cut too many trees, plugged rivers with logs, built dams and created fire hazards. The big mines dumped toxic chemicals and tailings and polluted both land and water. Ontario's north had become a land where greed and avarice ruled, and anyone objecting or fomenting strike action was handcuffed and given a free trip south. A greedy, heartless and rather lawless land until 1916, the north got a fresh start when weather finally intervened.

Forest fires are common in Ontario's north; they're a natural occurrence started by lightning and usually accommodated by forests. Tightly packed trees keep air from reaching fires and confine them to the underbrush until they encounter a natural firebreak, such as a river or meadow. In old growth forests, fires started by lightning are common and act like clean-up crews for the forest floor. Problems occur only when winds can reach flames through open spaces made by cutting too many trees. Small brush fires, sometimes deliberately set as part of brush-clearing projects and fanned by an unobstructed high wind, can quickly become a massive conflagration. The bottom line:

high winds and burning brush are a volatile mix and deadly to towns constructed entirely of wood and lacking firebreaks.

Nature's clean-up crew arrived at South Porcupine, a small mining settlement on Porcupine Lake across from Timmins, during the early hours of July 11, 1911. Spring had started early in 1911, the rains were long gone and the forests had turned dry as dust. Small brush fires had burned in the area for weeks, and fearful people longed for rain. That morning, miners at the South Porcupine settlement woke to the sounds of wind thrashing pines and a passing freight train. Never having heard a train in the area, the miners wondered if the railway had somehow run in a spur line. Half dressed, tin breakfast plates in hand, the miners trotted from the bunkhouses to have a look.

What they saw froze them body and soul—a wall of flame 30 metres high and making a noise louder than 10 freight trains. Tin plates scrunched under foot as screaming miners ran for their belongings, while some, wiser to the ways of the north, headed for the lake. Where they ran was incidental; the fire got them all. During the night, a descending subarctic storm system had brought down fast-moving cold, dry air and fanned all those small brush fires into one immense inferno—a firestorm.

The conflagration overran South Porcupine and burned every mining camp between Dome and Whitney townships. Then the wind direction became variable, and gusts fanned the flames higher and farther. By late afternoon, they could see the smoke and flames in North Bay, 300 kilometres to the south. The next day, having consumed an area 500 kilometres north and 100 kilometres east and west of South Porcupine, the fire destroyed the only way out—the rail junction at Cochrane.

In Timmins, the fire chewed the edges of town and swept past, but a change in wind direction brought it back to engulf most of the buildings as well as a railcar loaded with dynamite. The horrendous blast blew out a 10-metre-deep crater, which created a spring that, strangely enough, would become an impetus for rebuilding, because it offered the new town a constant water supply.

By the morning of the third day (July 13), with most fires burned out, roughly 2000 square kilometres of forest lay in smoking ruin, and the survivors had nothing but the singed clothing on their backs. More than 1000 people sought shelter in railroad boxcars, burned buildings and in Golden City (later Porcupine), a small settlement with a firebreak, thanks to a previous blaze. How many died is unknown; officials claimed 70 odd, but it was probably in the hundreds. The fire should have been a wake-up call for Ontario to do something to curtail brush burning. It wasn't, and they didn't.

Humanity's destiny seems to be to keep repeating mistakes until perfected, and perfection arrived on July 29, 1916. It was a complete repeat of the 1911 fire. It began in exactly the same manner—logger's brush fires spread by high winds. It burned just as long and destroyed just as much but took more than 200 lives.

The Matheson Fire (named after the town where it began) was a true wake-up call for the Ontario government. They immediately passed legislation to prevent yet another fiery repeat and sent along a corps of OPP constables to enforce firebreaks, oversee construction and issue brush-burning permits. Ontario's north prospered, and, in the 1920s a road was pushed through from North Bay to Cochrane. A few months later, the newly uniformed OPP detachment at Cochrane received their first shipment from the Queen's Printer in Toronto: a box of traffic tickets.

THE ANGEL OF LONG POINT

From the tip of Long Point on the north shore of Lake Erie to the American side (Long Point being directly across the lake from Erie, Pennsylvania) is a scant 40 kilometres of open water fringed on both sides by flats, shoals and sandbars. Dangerous enough, but thrusting into this strait is a here-today, gone-tomorrow sandbar extending from the tip of Long Point.

Dangerous waters for ships of old, but there existed an alternative to skirting Long Point; a channel cut through by a storm into Long Point Bay. Appropriately called the Old Cut, the channel saved precious time during storms and kept sailors out of the rigging to work sails. A ship's captain simply had to know where to aim his vessel, with no guessing allowed. Guessing wrong meant wrecked on a bar with all hands lost, but the crews of two unlucky ships had a saviour—Abigail Becker, the "Angel of Long Point."

In 1848, at age 17, Abigail Jackson from Townsend in Norfolk County, married widower Jeremiah Becker, a trapper from nearby Walsingham. He brought an instant family of five boys and a girl; Abigail eventually bore him three girls and five boys. The newlyweds moved to Long Point to be closer to Jeremiah's trap line. There, he built a tiny house from driftwood and washed-up timbers on the south side near the tip of the peninsula at a place called "the breakers."

Long Point suited Abigail, despite those lonely times when her husband travelled to Port Rowan to trade his furs. On many of his trading ventures, storms would rise and keep him gone for days, but, with a large family to look after, Abigail found purpose in the midst of loneliness.

One winter on the Point, and alone with her family, she risked her life in a blinding snowstorm to rescue the crew of a floundered schooner from freezing to death. Heroic enough, but, a few years later, it happened again.

On a blustery November morning in 1854, with her husband gone and huge waves pounding the shore and shaking the tiny cabin, Abigail ventured out to fetch water from a spring. Bucket in hand, head down against the driving snow, she happened to glance toward the lake and saw something that chilled her to the bone—a lifeboat. Dropping the bucket, she ran to the shore just as the boat smashed onto the beach. Thoughts raced through her mind: the boat was empty, how did it get here, where did it come from?

Sensing trouble, Abigail hitched up her skirts and ran along the beach toward the tip of Long Point. Running head down through driving snow, Abigail kept glancing up but saw nothing, until suddenly there it was—a ship, floundering on a bar and broadside to the waves. It was a real ship, like the ones she had seen in picture books, a three-masted schooner lying only 50 metres from shore.

Looking up, she could barely see the tops of the masts, and, when something moved in the rigging, she almost fainted. Men were up there. She counted eight, and she knew instinctively what had happened; fearful of waves and powerful currents, the crew had tied themselves to the rigging to wait for morning. But now that morning had arrived, they were too weak to move. She had to do something, but she was alone. What could she do? She thought: "Matches, do I have any? No, I had better get some and build a fire."

The schooner *Conductor,* en route from Amherstburg to Toronto via the Welland Canal with a cargo of 10,000 bushels of wheat, looked to be in dire straights. A stout ship, she was captained by Robert Hackett, a seasoned veteran who never took chances. But on that night, overtaken by a furious blizzard, he had decided to shelter in Long Point's inner bay and made a fatal error.

He was on a perfect tack for the Old Cut; one turn and his ship would be safe in the channel. He could see the red light kept burning beside the Cut, but the wind suddenly picked up. Blinded by driving snow, he could no longer see the light and decided to forego the Cut and make his way around the point. Some time later, still unable to see in the blinding snow, he guessed his ship had gone far enough east to make the turn and round the point. A few minutes later, *Conductor* went aground on a sandbar, shook herself free and grounded again, slewing broadside to the waves.

Knowing his ship was finished, Captain Hackett sought to get his crew to safety. One tender, slung over the side, had immediately disappeared into the maelstrom—a useless endeavour. It was not far to shore, only 50 or 60 metres, but monstrous waves and treacherous currents made swimming at night unthinkable. In the morning, with light to see, he and his crew would have a better chance. Turning to his seven anxious crew members, he yelled, "Into the rigging, lads. Tie yourselves good, now—and pray she hangs fast till first light."

Tied securely, Captain Hackett and his crew endured hours of freezing cold and wind, while the good ship *Conductor* steadily lost bits and pieces to the mighty waves. The captain had all but given up hope, when a nearby seaman caught his attention. He knew the young man was yelling something, but the wind made it impossible to hear. The seaman pointed down at the beach, and Captain Hackett saw a woman, waving her arms. Help had arrived, but in such a desolate, God forsaken place, how was that possible?

Hope rose, but when he saw that the woman was alone, it fell again. She waved again, and he waved back, but then she pointed down the beach and ran off. Hackett yelled for her to stop, but she kept running, her arms filled with soggy skirt. A sudden lurch of the ship set the masts to swaying and shook the rigging like the strands of a spider's web. Captain Hackett looked at his crew and began to recite the Lord's Prayer.

Abigail ran without looking back—even over the scream of the wind she could hear timbers snapping. The ship was breaking up—she would have to work quickly. Bursting into the cabin, she informed her oldest children of the wreck and filled their arms with cooking pots and blankets. Returning to the wreck, they gathered driftwood and lit a fire under the cooking pots. Then she turned and, braving the huge crashing waves, waded in up to her waist.

Abigail Becker could not swim—not a stroke—and staying on her feet to shout encouragement to those trapped sailors required an agility honed by hard years on the Point. "Move," she yelled. "Swim! I will help you!"

Looking down, Captain Hackett must have thought Abigail deranged. The tiny woman could hardly keep her footing in the monstrous waves. He looked over at his men and exclaimed: "If we stay, we are dead for sure." He quickly untied himself and, climbing down from the rigging, motioned his men to follow him down. On the wildly tilting deck, he kicked off his boots, removed his great coat, and yelled, "If I make shore successfully the rest of you follow, one at a time."

Hackett threw himself into the raging water and struck out for shore. Weakened from hours in the rigging made the going tough, but he never lost sight of the angel in the water. Twice he came within a few feet of her, only to be snatched away by waves that filled his eyes with sand—so much sand he could no longer see his angel. He could no longer fight, he was done, finished like his ship. Then he heard a whisper in his ear: "It's all right. I have you."

After dragging the captain to the warmth of the fire, Abigail coaxed all but one of the remaining crew into the water, pulling each to shore, where her children waited with a blazing fire, hot coffee and warm blankets. That lone crewmember, the ship's cook, too frozen by fear to move, refused all pleas to enter the water. The next morning, Captain Hackett and his crew fashioned a crude raft from broken timbers and poled out to the wreck, where they found the cook near death but still alive.

Abigail Becker's eight saved souls spread the tale of her heroic feat far and wide. To lake sailors she was the "Angel of Long Point," and they honoured her with a banquet in Buffalo, New York, where they presented her with a reward of $500. A few months later, a congratulatory letter from Queen Victoria arrived, along with a gift of £50.

When it was determined that three of the schooner's crew were American citizens, she received a solid gold, engraved medal from the Benevolent Life Saving Association of New York. The medal is inscribed: "Presented in May, 1857 to

Abigail Becker of Long Point, Lake Erie Canada West, for extraordinary resolution, humanity and courage in rescuing from impending death the crew of the schooner *'Conductor'* lost November 1854." On the reverse was engraved a picture of a schooner foundering in breakers on a beach. On the beach is a fire surrounded by people, and in the background the tiny cabin, home to Abigail Becker, the Angel of Long Point.

ONTARIO WEATHER STORIES

HAT TRICKS

Nothing shaped the early history of Ontario like the ubiquitous, weather-beater hat. At the dawn of civilization, humanity wore animal skins to keep warm and utilized the skulls as headdresses to denote tribe and rank, but they smelled awful and broke easily. Along came metal, and the animal skulls morphed into helmets with a feathered plume to denote nation and rank. They looked great but were heavy and cumbersome and could deafen the wearer in a rainstorm.

Then came the occasional peaceful interlude, and that heavy metal helmet saw replacement by all manner of millinery creations constructed from wool, flax or cotton. Of those, tried-and-true felt made from wool found the most favour. Felt made good hats that served a purpose in identifying nationality and rank, but it was sorely wanting in two basic utilitarian aspects. Wool insulates, which makes it an undesirable headpiece in hot weather. And it absorbs water, so a driving rain will soon turn it into soggy mush.

Not until about the middle of the 14th century did Dutch tanners discover that they could use waterproof animal fur to create felt, a process that required stripping the hair shafts from the hide and brushing it with a solution of nitrate of mercury—from whence the term "mad as a hatter" originates; breathing mercury fumes often drove the tanners insane. The tanner's fur of necessity was rabbit, but their choice was the water-repelling European beaver, because the end of each hair shaft contained a tiny hook that made felting easier and produced a finer product.

What a revelation, especially to the more than 15,000 English milliners and their minions. The beaver set their world on fire—finally, a felt that would protect from incessant rain and not turn to mush. Thousands of trappers turned on the European beaver so vigorously that, by the early 16th century, they were all but extinct.

European tanners tried desperately to find a satisfactory replacement, but nothing worked, and they feared a poor-quality product might find them in stocks or hanging from yardarms. Dark times, indeed, for European milliners and tanners, but then they got an unexpected break.

In 1532, a Frenchman from Brittany named Jacques Cartier returned from a voyage to the New World with his holds full of what he thought were mundane treasures: corn, beans and fur from strange animals. Among those furs in his hold were a fair number of not-so-strange furs…beaver pelts. One can only imagine the speed with which the milliners reached Cartier's dockside. Cartier's beavers made better felt than the European species, which started a race for more.

English, Dutch, French and Spanish trading companies fell over themselves in a bid to stake out New World trading grounds, and all manner of political intrigue ensued with the indigenous tribes. Europeans wanted beaver, and local inhabitants were happy to oblige. Problem was, the Europeans wanted a lot more beaver than the locals could supply, which necessitated the establishment of trading companies with outposts in Canada.

That worked for a while, until the indigenous tribes got out-of-sorts for being cut out of what they considered their domain. Native peoples turned hostile toward some Europeans and made deals with others. The Iroquois sided with the English; Ontario's Native peoples, the Tionontati, Algonquin and Huron nations, traded with the French. In 1630, the situation boiled up into what history refers to as the Beaver Wars (also called the Iroquois Wars or the French and Iroquois Wars).

This war lasted off and on until the turn of the century and caused the depletion of eastern Canada's beaver population. It almost resulted in the extinction of Ontario's Tionontati and Huron tribes by the Iroquois nations. With Canada East exhausted of beaver, the war moved to Canada West, or what is now Ontario, and resulted in the construction of English forts in the Great Lakes area to ensure a continuous supply of beaver fur to make those weather-beating hats.

History is weird, as is England's weather. Without their constant rainfall, the English might not have needed a waterproof hat, and Ontario residents might be speaking French.

ONTARIO WEATHER STORIES

THE RESCUE

Bob King woke on the morning of October 29, 1996, knowing a bad weather day was in the works. The sparrows in the tree outside his window always chirped, except in turning weather. On his way to shower, Bob glanced out the window and saw the sun rising in a clear sky. The morning looked fine, and he thought maybe the sparrows got it wrong.

Weather was important to Bob. As a Chief Coxswain of a Canadian Coast Guard vessel based in Thunder Bay, he captained a 14-metre patrol rescue boat that stood duty in all weather.

Driving through the city to the Thunder Bay docks, Bob kept the window down, soaking up autumn sunshine, because there would not be many more nice days before winter blew its ugly horn. He hated winters. His boat, the *Westfort,* was not designed for cold weather but stood ready all the same. Not much to do in winters, when pleasure boats were off the water, but the freighters came through in all weather, and he and his crew were there should an emergency arise.

Bob pulled into his parking spot, and, while walking toward the dock, he noticed his fine day had undergone a change. Hazy now, and he could hardly see Lake Superior for mist. So much for an easy day, he thought, and hopped onboard the *Westfort*. The smell of fresh-brewed coffee with eggs and bacon meant Inga Thorsteinson, one of his crew members, was already aboard. Inga made a good breakfast that neither he nor Willy Trognitz, his other crewman, would miss for any reason. After breakfast, the three busied themselves cleaning ship, while Bob's nice day progressively worsened; the hazy sky turned a greenish colour, and a breeze kicked in from the northeast.

In the same harbour, but at a different dock, the 23-metre tugboat *Glenada* had already undergone a stem to stern cleaning. Her captain, Gerry Dawson, and his engineer, Jack Olson, had come aboard early to prepare for a day of foul weather. Jack had 40 years of experience on Lake Superior and knew yesterday that the weather would turn ugly. Sure enough, a wind had started up from the northeast. Not that Jack was overly concerned; he spent all his time in the engine room and hardly ever experienced topside weather.

Getting cold up here, he thought, and motioned to Dawson that he was going below to his engine room. He knew without being told that it would be a sleepover. A blow was coming, and that would slow the arrival of freighters and back up the grain elevators: Cargill already had one waiting and the Richardson elevator two. This would be a long day and night with an early start in the morning.

At another dock, across the lake in another country—at a place called Grand Portage Bay, Minnesota—retired army major Dana Kollars and his Korean-born wife ChunAe stood looking out at their life's ambition, the 34-metre tour boat *Grampa Woo*. Named for his wife's grandfather and constructed entirely of aluminum, *Grampa Woo* had once served time as an oil rig work boat but was imminently more suited to the task of ferrying summer tourists around the islands of Lake Superior.

In a few days, once the new propellers had arrived and been installed, Dana and ChunAe would head south and take the craft to the Mississippi. They should have been underway days ago, but the props had yet to arrive. It made Dana uncomfortable to see his beloved *Grampa Woo* stripped of props and defenceless, but the weatherman had promised fair weather, and she had four anchor lines out. She was secure, the weather was fine—what could to happen to her?

Night fell on the headwaters of Lake Superior, and the lights of Thunder Bay shone bright against the night sky. Below deck in his bunk, Bob King watched them out a porthole for a few minutes before falling asleep. Over on the tug *Glenada,* Gerry Dawson and Jack Olson began a game of hearts that continued well past midnight and ended only when Jack's coffee cup tipped and rolled off the table. Jack hopped off his chair, walked to the ship's barometer and gave it a tap. He said nothing, just shook his head and headed for the engine room.

Only then did Gerry notice the creaking noise from the *Glenada*'s mooring lines, as the vessel rose and fell with the water. He got up, tapped the barometer and ran up the stairs to check the big barometer in the wheelhouse. It read the same, as low as he had seen it register. That meant trouble, and trouble is a tugboat's bread and butter. Gerry shouted for his engineer to get some sleep now, because they probably would not be getting any for some time.

Morning dawned the colour of green cement, and up on the *Glenada*'s bridge, Captain Dawson stood braced against a driving wind, scanning the outer harbour through binoculars. Jack stepped off the outside ladder carrying two cups of coffee, a trick that took 40 years to master—if a sailor lived that long. Jack handed a coffee to the Captain and shook his head at the offered binoculars. He didn't need a look to know port traffic would be standing down. Wind must be blowing 40 knots, he thought, and its only 6:30. By midmorning, it should be topping out at 60 knots. He took a sip of his coffee and scowled, cold already. "Gonna snow, boss," he said, as he headed for his toasty engine room.

Over on the *Westfort,* Inga had started breakfast and was kicking at the bulkhead to rouse her shipmates. Bob King awoke with a start, climbed from his bunk and stood staring out the porthole, smiling. Not even a fool would venture out in that, he thought, and no fools on the water meant an easy day for him and his crew. Bob suddenly realized he was famished and hoped Inga was whipping up a mess of pancakes and sausage.

At his home in Beaver Bay, Minnesota, Dana Kollars was up well before dawn and had headed out on his two-hour drive to Grand Portage Bay. *Grampa Woo* lay at anchor, tethered to a two-ton weight, so he should have nothing to worry about in any weather. Racing into Grand Portage Bay, Dana sped through town and skidded to a stop at Voyageur's Marina, just as the wind tore the roof off a small shed. Dana leapt from his car, ran around the marina building and onto the dock. Halfway to the end, his full run slowed to a walk. There was *Grampa Woo,* taking the waves like a trooper.

Dana breathed a lot easier, and when he spotted his deck hand, Robin Sivill, he even managed a grin. He told Robin to get the Zodiac tender ready and loped off into the marina to have a few words with the manager. It was nice and warm inside, and he had just begun talking, when Robin threw open the marina door and yelled, "*Grampa Woo* is moving!"

The pair leapt into the Zodiac and, with Robin at the controls, flew across the water. In minutes, they were onboard the *Grampa Woo,* throwing out extra anchors. Dana held his breath, hoping they would help, but no, they were not holding. *Grampa Woo* moved steadily toward open water, dragging a tremendous weight. Suddenly, the main tether snapped, and *Grampa Woo* bucked like a horse, throwing Dana onto the deck. Freed from its mooring and caught by the wind, the ship drifted out of the harbour and into Lake Superior.

Onto his feet again, Dana yelled, "Start the engines, Robin. We'll be needing the radio." He headed for a doorway while mentally assessing the situation. They were caught by a northeast gale blowing 50 or 60 knots: what to do…what to do? The sea anchor—it will slow us down and get *Grampa Woo* headed into the wind. Opening the door, Dana ducked inside to rig the sea anchor, but not before he saw something that chilled him to the bone: snow.

The sea anchor worked like a dream. *Grampa Woo* turned head into the wind and slowed to a bouncing crawl. Gasping from the exertion, Dana paused to catch his breath and saw a ship, a passing freighter, through the swirling snow. Running to the wheelhouse, he switched on the VHF radio and hailed the ship. Nothing. He tried again, and this time got an answer from the *Walter J. McCarthy,* a 1000-foot ore carrier out of Duluth, Minnesota. Dana quickly explained their predicament and received a "hang on, we're coming." It took the big ore carrier two passes, but she finally got her bulk between *Grampa Woo* and the wind.

The captain of the ore carrier then called over on the VHF told them to get aboard the *McCarthy*. But Dana would not abandon his dream, and Robin refused to even discuss leaving, so their only option was a tow. The *Walter J. McCarthy* sent down a line, the pair managed to secure it, and off they went behind the *McCarthy*, bobbing up and down on 10-foot waves like a giant yoyo.

Several hours later, just before dark and with weather conditions deteriorating rapidly, they passed into Canadian waters close to a string of rocky islands that guard the harbour of Thunder Bay. Captain Dana Kollars got on the radio to Thunder Bay Marine Services and asked for a tugboat to meet them by Pie Island, one of the string he knew was closest to Thunder Bay harbour. He then called the Canadian Coast Guard and explained the situation. He got a promise they would send a rescue boat and, for a few brief moments, Dana saw a light at the end of the tunnel. He could almost see Pie

Island, and the big ore carrier had begun a turn to Thunder Bay, but then the worst happened—the towline snapped. *Grampa Woo* suddenly lost speed and wrenched hard to starboard into a massive wave.

Captain Gerry Dawson on the *Glenada* got the call from Marine Services at about the same time as Chief Coxswain Bob King on the *Westfort*, and both used the same string of expletives to notify their crews. Engines started, the tug *Glenada* and the Canadian Coast Guard ship *Westfort* steamed full ahead, in a blinding snowstorm, into the dark, heaving waters of Lake Superior. Pie Island was not far as the crow flies, but that night, keeping a true course was impossible. To make matters worse, the *Westfort* was icing up at a dangerous rate. Bob thought about the safety of his boat and crew and considered turning back, but the two men onboard the *Grampa Woo* were in dire straights, so he had no option but to steam onwards.

He could see the *Glenada* in the lee of Pie Island, waiting and taking a terrible beating. Leaning over the wheel and peering into the darkness, Bob saw the lights of the ore carrier and the *Grampa Woo*. Only something wasn't right, they were getting farther apart. Bob knew what had happened and swore— the tow cable had snapped. He got onto the radio to the *Glenada* to apprise Captain Dawson of the situation and received a "We see it, and we're on it."

Gerry nosed the *Glenada* out from the lee of Pie Island and ran smack into a 12-metre wave that nearly swallowed his ship. He looked for the Coast Guard vessel and saw her roll 90 degrees and almost not right herself. Aware of her icing problems, he knew another giant wave would put the Coast Guard ship on the bottom. He had to get this over quickly, and he yelled for one of his deckhands to have a cable ready. In a few minutes, the deckhand, Jim Harding, entered the wheelhouse with bad news. All the cable was ice covered and frozen solid. Then he added that the wind speed was clocking more than 140 kilometres per hour.

Onboard the *Grampa Woo,* Dana and Robin were hanging on to this and that and watching the *Glenada* circle. Waves were striking *Grampa Woo* broadside and doing damage, and Dana wished that the *Glenada* would hurry and do something. That something turned out to be a nylon line thrown three times from the *Glenada* and finally grabbed and secured to the bow by Robin. Hope rose, but fell in minutes when the line snapped; the *Grampa Woo* was on her own again. Then Dana heard the worst over the radio from *Glenada's* captain. "The towing cables are all frozen, that was all we had. We are going to try to get you off now. Get yourselves ready."

From the ice-covered wheelhouse of the Coast Guard boat, Chief Coxswain Bob King watched the *Glenada* manoeuvre this way and that, trying to align with the *Grampa Woo* so both would be in the same trough of the giant waves. Finally, after a dozen tries, Bob saw the *Glenada* smack into *Grampa Woo's* side and stay there long enough for two figures to throw themselves onto the *Glenada* and be caught by a third. Mission accomplished, thought Bob, and he immediately turned to the problem of saving his own ship. There was no way the *Westfort* would survive another large wave, and they would surely encounter one or two. Then the radio crackled, and he heard the calm voice of *Glenada's* captain say, "Follow us. I know a safe spot."

The safe spot was a gravel beach in the lee of an island east of Thunder Bay. Bob watched the *Glenada* nose onto the beach, thought "what the heck" and followed suit. The two craft waited out the storm together—engines running, lights blazing and the stereo cranked up to drown the awful howling of the wind. The party, if you could call it that, lasted for almost three days. When the seas finally calmed, the two ships sailed back to Thunder Bay, where the city gave them a hero's welcome.

Two years later, the crews of the *Westfort* and the *Glenada* were awarded the Governor General's Medal of Bravery, "for acts of bravery in hazardous circumstances," for their role in the rescue. A year after that, Jim Harding, who risked his life on the icy bow of the *Glenada* to grab Dana Kollars and Robin Sivill as they leapt off *Grampa Woo,* had his medal upgraded to the Star of Courage, "for acts of conspicuous courage in circumstances of great peril."

All's well that ends well, but not for the *Grampa Woo.* Smashed onto the rocks during the storm, she suffered the ultimate indignity of being classified as rubbish. Dana Kollars was forced to have her torn off the rocks and sunk in deep water. On a brighter note, Dana and his wife have a new and bigger *Grampa Woo.*

A BIBLE STORY REDUX

The Creator saw that all was not well with his garden and sought out the prodigy of Noah, who turned out to be a carpenter who dwelt in a place called London, in a province of Canada called Ontario.

"Noah, it is the Lord thy God speaking."

Noah Smith, 36, recently separated and unaccustomed to being addressed by a burning bush in his living room, tried desperately to extinguish the flames with pots of water from his kitchen sink. Unable to do so, he fell onto his knees, exhausted.

"Noah," said the Lord. "I made thy relative a promise to never again cleanse the world with water, but it went along with the understanding that my garden be treated in a respectful way. This has not happened; the garden is so befouled, I have decided to rescind that promise."

"Befouled," said Noah, staring through unbelieving eyes. "How's that?"

Employing infinite patience, the Lord explained. "Mankind has made unclean the essence of my labour and turned paradise into a cesspool. You have cut all the trees, turned my whales into snack food and made the waters undrinkable."

Noah, who did not like sushi and drank only bottled water, nodded agreement and jumped when a large roll of paper flew out from the burning bush.

"Those are plans for the new ark, Noah. When completed, ye shall gather the creatures two by two and prepare for the flood. I shall hold off the rains for six months."

Poof, the burning bush disappeared, leaving Noah to clean up the mess.

And it came to pass, in pouring rain, that the Lord once again spoke to Noah from the burning bush in a loud and angry voice. "Noah! Where is the ark? The flood has already begun."

Noah, falling onto his knees, cried out, "Forgive me, Lord. I laid the keel, but the building department issued a stop-work order, and my neighbour got an injunction forbidding me to build in my backyard. I am still before the Developmental Appeals Board for a decision on the matter."

"My animals," sayeth the Lord. "Where are the animals I charged you with saving?"

"Forgive me, Lord. When I began to gather them, I was beset by animal rights groups, and the animals were let go. Then Environment Canada decided I needed an Environmental Impact Statement and issued me a cease-and-desist order. Then my building crew was dismissed, until the Human Rights Commission decides how many minorities I must hire to build the ark."

"But Noah, I see no sign of your labour."

"Seized by Canada Customs and Revenue after they decided I was guilty of conspiring to smuggle animals out of the country. Forgive me, Lord, it will take me 10 years to build your ark."

Up to his knees in mud and expecting wrath, Noah was surprised when the sky suddenly cleared, allowing the Sun to shine upon the Earth. Looking into the burning bush, Noah said, "Does this mean you won't be destroying the world, Lord?'

With the flames of the bush petering out, Noah heard the Creator chuckle and say: "My hand will not be necessary to destroy the world, Noah."

Conclusions

Isn't it interesting that the same people who laugh at science fiction listen to weather forecasts and economists?

–Kelvin R. Throop, a.k.a. R.A.J. Philips, novelist

CONCLUSIONS

WEATHER THOUGHTS

Pay Attention

If you have learned little from the preceding chapters, not to worry, because you're in good company. Weather experts abound, but the truth is, not many people know much about the subject. One can usually count on experts for a daily forecast, but you can have that information by accessing Environment Canada's radar website and calling your sister in the next county. "Hey Martha, is it raining where you are? I want to take the dog for a long walk."

CONCLUSIONS

In Ontario, rain is sometimes the least of your concerns. You can take a raincoat or umbrella for some precipitation protection on your walk, but there's not a lot you can do if Martha's rain heralds the coming of that mysterious atmospheric behemoth called a supercell. Scientists know a lot about supercells, but many unanswered questions remain. Why do only some spawn tornadoes? Why do only some produce hail? Why do some dump small hailstones and others large? How can tornadoes form in different sections of the supercell cloud? How can dry-air supercells spawn tornadoes usually associated with wet-air cells? Does cloud seeding affect tornadoes spawning from supercells? Nobody knows, and there are lots more unanswered questions, but one thing is certain: be especially leery of light rain and overcast skies, because foul weather is the colour of cement and good for hiding monsters.

But what the heck. Despite some qualms, you throw caution to the wind and head off behind a dog that seems inexplicably anxious. His angst is only inexplicable to you, because small mammals, such as dogs, raccoons, possums, squirrels and coyotes, know when a dangerous storm is brewing. The great earthquake that shattered Indonesia in 2004 created a monster tsunami that killed 100,000 people but few wild animals. Somehow, the wild critters had sensed the coming wave and made a beeline to higher ground.

Truth is, pretty much all of nature knows when a dangerous storm is imminent. Bugs fly low, birds not at all, small mammals remain in their holes, trees turn their leaves and flowers close for the duration. Alas, humans can only stand and stare when the curtain rises to reveal a supercell, when thunderstorms, lightning, tornadoes, derechos, tomato-sized hail, falling trees, flying limbs, downed power lines and flash floods can be expected. "Supercells R Us" has a home base in Ontario, and they throw around catastrophe like drunken sailors on New Year's Eve. Be alert to their dangers, and if in the morning you have no idea what lurks on the horizon, throw the dog out the door and wait. If danger is out there, the pooch will finish up quick and return in a flash.

Better yet, know your area's storm history, especially if you are a new arrival. If you're visiting Ontario, especially in the southwestern regions, keep in mind that sunny mornings don't necessarily translate into sunny afternoons. Stay alert for supercell storms, because you might encounter a surprise and find your stay more permanent than planned.

Most storms will not smash your house and vehicle; the majority simply provide free water for your lawn and a wash-down for the road. Rain cools, cleans the air and replenishes the water table. Storms are beneficial and necessary, but don't turn your back on any, because all are capable of fatal surprise.

Questions anyone?

Run or Walk?

How about a question that has confounded civilization since people began wearing stainable fabrics? The famous and eternally argued query: is it better to run or walk in a rainstorm? A conundrum that no one cared about until the wraparound sheet became a 100-drachma extravagance. "Run, you fool! Do you want the rain to stain your new toga?" Some ran, while others, convinced forward motion would only increase collision with raindrops, insisted on walking to shelter. Great debates and arguments ensued and continue to this very day.

Who is right? The question, old as linen, was only recently addressed via an experiment by American climatologists. They convinced groups of students to wear the same weight of cotton clothing while walking or running through rainstorms. After each soggy session, the climatologists weighed each student's

CONCLUSIONS

clothing and the heft differential of runners verses walkers finally put a solution to the question.

Answer is…'tis better by far to run. Much better, in fact—running to a shelter will leave you 40-percent drier than walking. A good tip. Not nearly as good as, say, stay off golf-course fairways in stormy weather—but still good. One tip can save you a trip to the cleaners, and the other might save your family a trip to the cemetery.

Weird weather kills, and in Ontario that weather is everywhere, so be aware and don't shelter from storms under trees or highway overpasses, and do not dance in the rain. Do pick or build a room to protect you and yours from tornadoes (remember Barrie—May 1, 1985), and please do your bit to help our planet endure human infestation, the cause of unnatural climate change.

CLIMATE-CHANGE THOUGHTS

It's Not All Doom and Gloom

In 25 years, Asia's population will double, its forests will be gone, numerous cities with 50-million-plus residents will spew toxic contaminants onto exhausted agricultural lands and millions will starve. It's inevitable, and already actions in Asia are beginning to cause drastic changes to Ontario's climate—increased airborne particulates and ozone, rising seasonal temperatures, forest fires burning out of control and rainfall contaminated by heavy metals. Today's weird weather in Ontario might be enviable in 25 years, and weather then will be our fault, because we worried about symptoms, rather than addressing the chronic disease of global overpopulation.

Nuclear power, garbage incineration, coal, hydro and geothermal energy have become political agendas for Ontario's green gangs. A sad situation, because time wasted bickering and fighting over these forms of cheap energy would be better used finding solutions for the problems each creates. All the energy Ontario needs for centuries to come is there for the taking—all we need to do is connect the dots.

Nuclear energy is ideal. Find a way to safely dispose of the spent fuel, and Ontario residents could continue to have cheap electricity for centuries. Nothing is impossible—there has to be a way to solve this problem, but it will need political will and massive funding. Coal and garbage can be used, though scrubber technology needs perfecting to the point of discharging perfectly clean air.

Hydro-generated electricity is already a clean source of energy, and the potential for more is almost unlimited in a province

CONCLUSIONS

of countless lakes and rivers. Ontario has more than 1000 sites suitable for the installation of small, low-environmental-impact hydroelectric generators and another 1000 that would require careful installation so as not to impact wildlife. But take care: damming rivers to produce clean hydroelectricity, without thought to the submerged ecosystems that result, is insane.

Ontario is a province of islands in an ocean of unutilized fresh water that every year is looking more attractive to our parched neighbours south of the border. A few hundred kilometres into northern Ontario, one encounters the Arctic watershed, where great rivers, such as the Mattagami, Groundhog, Missinaibi and Chapleau, flow northward without apparent benefit to humanity. Wasted water, better utilized to quench the rising thirst of a few million acres of America's corn-growing states. Self-sufficiency in fuel is America's goal, and corn, a biofuel crop that needs an endless supply of fresh water, is part of their game plan. *You* don't need that water; it's going north to nowhere, and who will even notice if it's siphoned off to the U.S.? Well, other than the lunacy of turning edible crops into fuel, clearing huge tracts of forest to divert rivers to grow biofuel causes weather problems worldwide.

Geothermal energy opportunities are everywhere in Ontario; the brand-new University of Ontario Institute of Technology near Pickering drilled eight bore holes deep into the ground and now produces enough power to cool and heat the equivalent of 1000 homes. The heat is there, under the ground and awaiting utilization in a completely ecofriendly manner.

Although other, already ecofriendly forms of power generation are currently in vogue—wind turbines, solar power and biomass—all are costly, and large-scale installations make little economic sense. The electrical output of wind and solar farms depends on too many variables and requires backup by standard power-generating plants that are constantly ramping up and down, depending on the variables. This slowing down and speeding up actually creates more emissions than a standard generating plant running at full capacity.

Solar energy does have a future, but currently it's an unreliable energy source, with installations limited to rooftops and chunks of desert defended by groups wanting to protect moles, voles and burrowing owls. Solar energy also has an ugly side. To make the photon-converting panels requires heating vast amounts of sand to ultra-high temperatures to produce polysilicon. High temperatures require enormous amounts of energy, and, in China, where most polysilicon is manufactured, that means burning dirty coal. Polysilicon production also creates an extremely toxic waste called silicon tetrachloride that, evidence suggests, is being dumped into the Chinese countryside.

Thinking outside the box might secure the future of solar energy, because, sometime down the road, the moon will begin looking attractive to utility companies. Covering a portion of the lunar surface with solar cells and beaming generated power back to Earth must be an idea already creeping into the plans of those companies. A solar power array on the moon would mean almost-free, unlimited energy and could be a reality within five years—if not for global military spending.

If, somehow, the world could stop the insanity of military squander and get its population growth under control, our planet could be a Garden of Eden redux. Hey, it could all be a test: God with a stopwatch, seeing how long it takes for the lambs to stop turning his/her beloved trees and whales into charcoal and sushi. But as the watch ticks, the lambs still frolic and the Garden continues to decay.

On the brighter side, the watch *is* still ticking, and the big blue marble is still fixable. A task we better get on with,

CONCLUSIONS

because I have this sneaking suspicion that Earth was meant to be Heaven 101. If that's true, how will we explain the filth and infection to the Creator? We're behaving like puppies in a box—pooping, whining and eating everything in sight. Maybe that old adage "you reap what you sow" is not so old-fashioned, after all. Oh Lord, please be merciful!

NOTES ON SOURCES

Barrett, Harry B. *Lore and Legends of Long Point*. Toronto: Burns and MacEachern, 1977.

Cooper, Charles. *Rails to the Lakes*. Cheltenham: Boston Mills Press, 1980.

Dyson, Freeman J. *From Eros to Gaia*. New York: Pantheon Books, 1992.

Dyson, Freeman J. *Imagined Worlds*. Cambridge, Massachusetts: Harvard University Press, 1997.

Higley, Dahn D. *O.P.P, The History of the Ontario Provincial Police Force*. Toronto: The Queen's Printer, 1984.

Jensen, Derrick, and Draffan, George. *Strangely Like War*. White River Junction, Vermont: Chelsea Green, 2003.

Passfield, Robert W. *Building the Rideau Canal*. Toronto: Fitzhenry and Whiteside, 1982.

Stewart, Darryl. *Point Pelee*. Toronto: Burns and MacEachern, 1936.

Turner, Thomas P. *Weather Patterns and Phenomena*. New York: McGraw-Hill, 1994.

Wood, Richard A. *The Weather Almanac 10th edition*. New York: Gale Group, 2001.

Online Sources

Bruce Trail Conservancy: www.brucetrail.org

Environment Canada, Ontario: www.on.ec.gc.ca

Meteorological Service of Canada: www.msc-smc.ec.gc.ca

Ontario Weather Page: www.ontarioweather.com

ABOUT THE AUTHOR

A.H. Jackson

H. Jackson has always been fascinated by the skies above. A certified pilot, he flew his first plane at 14 and left home at 17 to continue his atmospheric exploration. Jackson believes that, in the twine of life, there are two special genes unique to humankind—hope and humour—and thinks we should all turn to the funny side of life in the face of adversity. He must have quite the sense of humour, then, since he's been struck by lightning five times!

When not writing about weird weather, Alan is a creator of worlds, a fiction writer. For children mostly, because they have imaginations unconstrained by reality. Can pigs fly? No, but in one of his books, *Growing Bob,* a pig talks, plots and saves mankind from becoming the bottom link of the food chain.

Alan lives in Toronto with a wife named M and a squirrel called Mommy. He is the author of *Weird Canadian Weather*, also from Blue Bike Books.

ABOUT THE ILLUSTRATORS

Peter Tyler

Peter is a recent graduate of the Vancouver Film School's Visual Art and Design and Classical animation programs. Though his ultimate passion is in filmmaking, he is also intent on developing his draftsmanship and storytelling, with the aim of using those skills in future filmic misadventures.

Roger Garcia

Roger Garcia is a self-taught artist with some formal training who specializes in cartooning and illustration. He is an immigrant from El Salvador, and during the last few years, his work has been primarily cartoons and editorial illustrations in pen and ink. Recently, he has started painting once more. Focusing on simplifying the human form, he uses a bright, minimal palette and as few elements as possible. His work can be seen in newspapers, magazines and promo material.

Roly Wood

Roly grew up in Indian River, Ontario. He has worked in Toronto as a freelance illustrator, and was also employed in the graphic design department of a landscape architecture firm specializing in themed retail and entertainment design. In 2004, he wrote and illustrated a historical comic book set in Lang Pioneer Village near Peterborough. Roly currently lives and works as a freelance illustrator in Calgary, Alberta, with his wife, Kerri, and their dog, Hank.

MORE TRIVIA FROM BLUE BIKE BOOKS...

BATHROOM BOOK OF ONTARIO HISTORY
Intriguing and Entertaining Facts about our Province's Past
by René Biberstein
From the Great Lakes to Hudson Bay, Ontario is a province rich in history—and some of that history is simply weird. Read about the fascinating and funny past of Canada's most populous province.

$9.95 • ISBN: 978-1-897278-16-1 • 5.25" x 8.25" • 168 pages

BATHROOM BOOK OF ONTARIO TRIVIA
Weird, Wacky and Wild
by René Biberstein
From Toronto to Kenora, Windsor to Fort Severn, Ontario is definitely an interesting place—home to two weather-predicting groundhogs, Wiarton Willie and Spanish Joe, and the world's longest gum-wrapper chain. Read more about the province that brought you BeaverTails and the Wonderbra in this easy-to-read collection of fun facts.

$9.95 • ISBN: 978-1-897278-03-1 • 5.25" x 8.25" • 168 pages

WEIRD ONTARIO PLACES
Humorous, Bizarre, Peculiar & Strange Locations & Attractions across the Province
by Dan de Figueiredo
Canada's most populous province may also boast some of the weirdest places in the country. This enjoyable collection features hundreds of odd locales and structures.

$9.95 • ISBN: 978-1-897278-07-9 • 5.25" x 8.25" • 168 pages

WEIRD CANADIAN WEATHER
by A.H. Jackson
This book covers everything from the Chinooks of the Prairies to the different types of snow crystals, flash floods, ice storms, droughts and everything in between. It includes statistics such as the warmest lake waters in Canada and where exactly the most rain fell in just one hour.

$14.95 • ISBN: 978-1-897278-39-0 • 5.25" x 8.25" • 224 pages

WEIRD CANADIAN PLACES
Humorous, Bizarre, Peculiar & Strange Locations & Attractions across the Nation
by Dan de Figueiredo
The Canadian landscape is home to some pretty odd sights—for example, the UFO landing pad in St. Paul, Alberta, the ice hotel in Québec City or Casa Loma, Canada's only castle. This book humorously inventories many real-estate oddities found across the country. Welcome to the True North, strange to see.

$9.95 • ISBN: 978-0-9739116-4-0 • 5.25" x 8.25" • 168 pages

Available from your local bookseller or by contacting the distributor,
Lone Pine Publishing
1-800-661-9017
www.lonepinepublishing.com